Heinemann Investigations in Biology

General Editor: S. M. Evans

Seashore and Sand dunes

Seashore and Sand dunes

S. M. Evans, BSc., PhD.
Senior Science Master, Archbishop Holgate's Grammar School, York

and

J. M. Hardy, BSc.
Assistant Biology Master, Archbishop Holgate's Grammar School, York

Heinemann Educational Books Ltd
London

Heinemann Educational Books Ltd

LONDON   MELBOURNE   TORONTO   AUCKLAND   EDINBURGH
JOHANNESBURG   SINGAPORE   HONG KONG
IBADAN   NAIROBI   NEW DELHI

ISBN 0 435 60280 2

Published by Heinemann Educational Books Ltd
48 Charles Street, London W1X 8AH
Printed in Great Britain by Richard Clay (The Chaucer Press), Ltd
Bungay, Suffolk

Preface

This book presents a series of investigations in the biology of the seashore and sand dunes which are considered to be suitable for both project- and class-work by sixth-formers, first-year undergraduates and students in colleges of education. With only one exception, they have been used by the authors in their own work with students so that they have been tested and are known to produce interesting results. In general, the investigations involve the use of well-known organisms, which are common and easy to identify. Furthermore, they require only simple apparatus and minimal laboratory facilities.

The large majority of investigations are based on published research, and it is hoped that they will introduce the student to the kinds of problems that have interested marine biologists and the ways they have devised to solve them. We have tried to include enough information about them for students to carry out each one successfully and to appreciate the difficulties involved in experimental work. At the same time we have not specified experimental techniques in too much detail. It seems to us that there is a danger that students, given too much information, stop thinking for themselves. This would be particularly unfortunate, because there can be no doubt that there are abundant opportunities for further research on the seashore and it is felt that this book would serve a valuable function if it stimulated students to form their own hypotheses and design experiments to test these for themselves. Surely there can be no better training in scientific method?

References to some of the more important papers are given at the end of each investigation, so that the interested student can find his way into the relevant literature and read more fully into work that attracts him. In most cases papers have been selected which are non-technical and could be read profitably by a sixth-former of average ability. It is, of course, particularly important to consult research papers when original work is contemplated. The necessary scientific journals can be found in university and specialist libraries, but can also be obtained from public libraries through the inter-library loans service.

However, this book is not intended to provide a complete course in the ecology of the seashore and sand dunes and, indeed, it would be unfortunate if work of the kind suggested here led to detailed investigations of a limited number of organisms by students who had little general knowledge of the habitats. To some extent this can be offset by reading, but we believe that the first necessity for a student visiting the shore is to familiarise himself with the common animals and plants that live there. Not that this needs to be an unduly lengthy or arduous task. Most common organisms can be identified from Dr. J. H. Barrett and Professor C. M. Yonge's excellent book, *Pocket Guide to the Sea Shore*, Collins, 1958, although specialist keys may be necessary for rarer organisms and reliably accurate identifications. References to suitable keys will be found in the book by Dr. N. B. Eales, *The Littoral Fauna of the British Isles*, Cambridge Univ. Press, 4th. edn., 1967.

We would like to acknowledge the help of numerous past pupils who have accompanied us on visits to the coast and have tried out these investigations. In particular, we would like to thank Mr. Grahame Patterson for his help with the section on parasites and commensals. We are most grateful to our colleague, Mr. R. C. V. Rose, for Plates 1 and 3–9 inclusive.

S.M.E.
J.M.H.

York, 1969

Contents

List of Plates

Acknowledgements

We would like to acknowledge permission from the following to redraw figures: G. Bell and Sons Ltd. (Figure 20); Blackwell Scientific Publications Ltd. (Figures 2, 3 and 5); Cambridge University Press (Figure 7); English Universities Press Ltd. (Figure 1); The Marine Biological Association (Figures 6 and 14); The Malacological Society of London (Figures 4 and 8); Methuen and Co. Ltd. (Figure 11); Pergamon Press Ltd. (Figure 19); and William Collins and Sons Ltd. (Figures 16 and 17).

1

Adaptations to life on the seashore

The seashore is usually considered to be the region between land and sea that is affected by the tides. However, because the highest and lowest limits are only occasionally covered or uncovered by the sea, it is not as clearly delineated as this definition might seem to imply. The reason is that the vertical range of the tide increases and decreases according to a lunar cycle, with the maximum (spring) tides (i.e. those with the greatest distance between high and low water) occurring at about fortnightly intervals (at full and new moons) and the minimum (neap) tides occurring between them. In fact, the differences between tides can be considerable; on the Pembrokeshire coast, for instance, the mean range of the spring tides is about 20 ft., compared with about 7 ft. for the neap tides. Actually, the situation is still more complex because an annual rhythm is superimposed on this lunar rhythm and all tides tend to be larger at the equinoxes (about March 21st and September 21st). The spring tides are particularly impressive at these times, and because such large areas are exposed at low water they provide unique opportunities for collecting organisms that are not usually uncovered by the tide.

The character of the shore may vary enormously from one place to another. It may, for example, slope gradually from high to low water so that the distance between the tide marks is great or, alternatively, there may be a steeper gradient so that the high and low tide marks are relatively close together. Similarly, the nature of the substratum may vary and, in different places, may consist of rock, sand, clay, shingle, mud or two or more of these types merged together. However, with the exceptions of shores of clean sand and shingle, the intertidal zone usually supports an astonishing number of organisms, although, as one might expect, different types of shore tend to have their own characteristic populations. Rocky shores, for example, which are often exposed to considerable wave action, are usually well populated by organisms which have some means of adhering firmly to rocks, such as limpets, barnacles and brown algae. On sandy and muddy shores attachment to the substratum is difficult, and burrowing animals, particularly worms and bivalve molluscs, tend to dominate the populations.

But although the seashore is so richly colonised, life there poses special problems. Intertidal organisms must, for example, be able to tolerate or avoid (e.g. by burrowing) the dangers associated with regular exposure to air. Perhaps the most serious of these is the risk of desiccation, because most animals and plants that live on the shore have moist surfaces and rapidly lose water by evaporation in air. There are also greater fluctuations in the temperature of the air than that of the sea, and during excessively hot or cold weather there is a danger of exposure to potentially lethal temperatures when the tide is out. Intertidal organisms must also be capable of tolerating the temporarily reduced salinities caused by rainfall, unless, like limpets, which clamp firmly down on rocks when wetted by freshwater, they have some means of avoiding them.

However, despite the difficulties of surviving this regular exposure to air, some organisms cannot survive without it. This is evidenced by the absence of many common shore-living organisms in rock pools. Barnacles, for example, tend not to occur in them, and there is often a sudden demarcation line at the edge of a pool where the growth of barnacles stops. Similarly, some algae, such as *Fucus spiralis*, cannot survive indefinite submersion in sea-water and only live if they are periodically exposed to the air. The reasons for this are not properly understood, but one suggestion is that the efficiency of photosynthesis may be impaired when plants are submerged because only diffuse light reaches them. It is also known that some algae, of which *Fucus spiralis* is one, discharge their gametes more effectively after a period of drying.

Conditions are not, of course, uniform on all parts of the shore. In fact, there is a progressive change from high water, where organisms are exposed to air for most of their lives, to low water, where they are only exposed to it for a few minutes when the tide is fully out. Not surprisingly, most organisms only appear to be able to live successfully under a limited range of these conditions, and as a result, most of them are confined to distinct zones on the shore (Figure 1). On many rocky shores, for example, the channelled wrack, *Pelvetia caniculata*, and the tiny periwinkle, *Littorina neritoides*, are characteristic organisms of the splash zone (i.e. the region above high water which is affected by spray but is not covered by the tide), but do not usually occur much lower down the shore. The serrated wrack, *Fucus serratus*, and the flat periwinkle, *Littorina obtusata*, are normally restricted to the mid-tide level, and the oar-weed, *Laminaria*, which is a large brown algae with strap-like fronds, and the beautiful blue-rayed limpet, *Patina pellucida*, are usually only found near and below low. water. Similar zonations occur on sandy and muddy shores.

Zonation of many of the common algae on the shore is often obvious, even to the casual observer, as a series of distinct belts of differing hues and colours, although the zonation of animals is usually less obvious. Quantitative estimates of zonation can be made by taking a line transect down the shore. The transect usually starts from the upper limit of the spring tide level, including the splash zone, and extends down to the low-water mark. A series of stations at intervals of 5, 10 or more metres, depending on the situation on the shore, should be taken along the transect. At each station the number of organisms present should be estimated by the quadrat method. On rocky shores a wire frame (1 m² is a suitable size) is placed on the substratum and the number of individuals of each species that are within it are counted. (For barnacles and some algae this is often impracticable, and an alternative is to estimate the percentage of the surface area that is covered by them.) On sandy and muddy shores the substratum must be sieved for accurate estimates of the

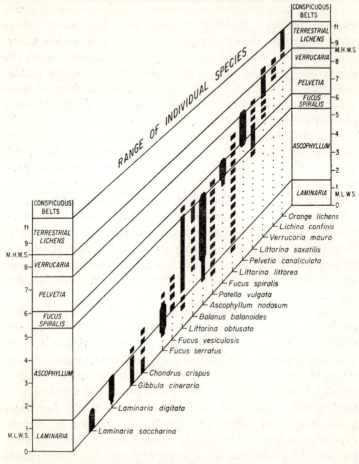

Figure 1. A diagrammatic representation of the zonation of some common animals and plants on a rocky shore. (After Lewis.)

numbers of organisms present. It is surprising how many burrowing animals are missed even by the most careful digging and sifting. Preferably the substratum from an area (quadrat) of between, say, 0·1 and 0·25 m² should be dug out and collected in a bucket or a larger container. Only small amounts of this can be washed through a sieve at a time. This is done by dunking and agitating the sieve and its contents in the nearest water (e.g. in a pool or at low water). A suitable sieve is about a foot to eighteen inches square with sides a foot deep. The bottom should be covered with perforated zinc or some other perforated material.

Recommended books for general reading
Lewis, J. R. (1964). *The Ecology of Rocky Shores*. English Universities Press Ltd.
Southward, A. J. (1965). *Life on the Sea-shore*. Heinemann Educational Books Ltd.: The Scholarship Series.
Yonge, C. M. (1949). *The Sea Shore*. Collins: New Naturalist Series.

Investigation 1: Resistance to desiccation in algae

Desiccation resulting from excessive water loss by evaporation in air is undoubtedly one of the most serious problems for both animals and plants on the seashore. In general, it is more important to organisms living high up the shore than those found nearer low water, because they are exposed to the air for longer, and several investigators have established relationships between the ability to tolerate and resist desiccation and the position normally occupied on the seashore. Isaac (1933; 1935) has shown that just such a relationship exists among the brown algae. The channelled wrack, *Pelvetia canaliculata*, which is found in the splash zone on rocky shores, can lose up to 60–80% of its total weight and still remain viable, whereas *Fucus serratus*, a mid-shore species, can only lose up to 40% and survive, and *Laminaria digitata*, which is normally only exposed at low tide, cannot tolerate losses of more than 20–30%.

Zaneveld (1937) made an interesting comparative study of four species of the Fucaceae: *Fucus spiralis*, *F. vesiculosus*, *F. serratus* and *Ascophyllum nodosum*. These usually have a well-defined zonation (Figure 2) and are adapted to living at different levels on the shore. Zaneveld found that those species living high up the shore, *F. spiralis* and *A. nodosum*, initially have a higher water

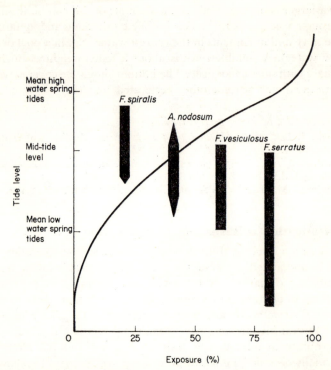

Figure 2. The zonation of four species of Fucaceae on the Leidam and Harsens in relation to tidal levels. The percentage amount of exposure is shown at different levels. (After Zaneveld.)

content than the lower-level species, and they lose water by evaporation more slowly (Figure 3).

(*a*) *The water content of algae and the loss of water in air*
Zaneveld's experiments can be repeated in the laboratory with the species of Fucaceae used by him and, if desired, various red and green algae from different levels on the shore; *Ulva* and *Enteromorpha*, for example, are good experimental subjects. The algae that are used should be kept in sea-water for an hour or so beforehand to ensure that they are fully hydrated.

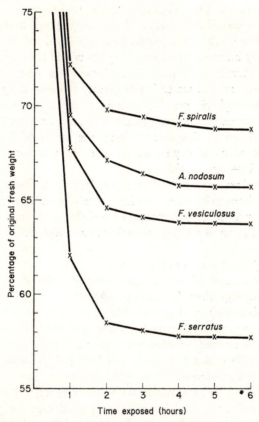

Figure 3. The rate of water loss, measured by decrease in fresh weight, in four species of Fucaceae after exposure to air. (After Zaneveld.)

The loss of weight that occurs from pieces of thallus in air can be used as an index of the amount of water that is evaporated from them. Several pieces of thallus of about equal sizes should be cut from the ends of healthy fronds and the cut ends dipped in olive oil or liquid paraffin to prevent water loss from them, taking care to prevent oil from reaching other surfaces of the plant. Before weighing, excess water should be removed from pieces of

B

thallus by dabbing them lightly with absorbent paper. Each piece should be weighed separately and then left exposed to the air in a dish or on non-absorbent paper. Reweighings should be made at half-hourly intervals for at least four hours and then the percentage loss in fresh weight plotted graphically against time for each species. The rate of water loss is greatly influenced by the prevailing atmospheric conditions and, as might be expected, is much slower on damp, humid days than on drier ones. It may be considered better, therefore, to keep the pieces of thallus in a desiccator above a drying agent, such as anhydrous calcium chloride, between weighings.

The total water content of pieces of algae can be determined by allowing them to become completely desiccated. They should be kept in dry air in a desiccator and reweighed until their weight is constant (i.e. until there is no further loss of water).

(b) *Absorption of water by algae*

Partly desiccated pieces of algae usually re-imbibe water extremely rapidly when they are brought into contact with it again in the laboratory, and presumably soon make good water loss when they are covered by the incoming tide. The rate of water uptake can be studied by submerging partially desiccated pieces of algae (totally desiccated pieces will be dead) in sea-water and measuring the increase in weight that takes place. Pieces of thallus should be prepared in the manner described above and left in air for, say, two hours. They should then be placed in sea-water and weighed at intervals of about five minutes. Before each weighing excess water should be removed from pieces of thallus by dabbing them with absorbent paper.

It is unlikely that losses of water by algae under natural conditions are as excessive as the results of laboratory experiments might seem to imply. One reason is that many of the Fucaceae and other seaweeds occur in dense clumps on the shore, and there is undoubtedly absorption of water by plants which are exposed to the air from unexposed plants. The importance of this absorption can be demonstrated by estimating the rate of water loss in pieces of thallus placed on top of, and in contact with, bunches of algae. This should be compared with similar pieces of thallus, as controls, that are placed on top of other plants but are separated from them by polythene sheeting so that water exchange cannot occur. The experiment works well if the covered algae are protected in deep dishes so that they do not themselves become desiccated. It should, of course, be borne in mind that this experimental situation is complex and that a zone of damp air is undoubtedly established around large bunches of algae. This almost certainly reduces water loss, both in this experiment and on the seashore.

(c) Cell-wall thickness

According to Zaneveld, a large amount of the water that is held by brown algae is contained in the unusually thick cell walls of these plants. He also found that there was a correlation between the thicknesses of the cell walls of different species and their position on the shore. The high level forms, which hold more water, have thicker cell walls than algae from lower down the shore (Table 1).

TABLE 1

The thicknesses of the cell walls of four species of Fucaceae. (After Zaneveld.)

	Cell-wall thickness (μ)
Fucus spiralis	$1 \cdot 47 \pm 0 \cdot 05$
Ascophyllum nodosum	$1 \cdot 02 \pm 0 \cdot 03$
Fucus vesiculosis	$0 \cdot 69 \pm 0 \cdot 09$
Fucus serratus	$0 \cdot 42 \pm 0 \cdot 03$

The thickness of cell walls of different species can be compared by cutting good sections of various algae and examining them microscopically. If a micrometer eyepiece is available actual measurements of the thicknesses can be made. For comparative purposes, the cells measured should be from the same part of the thallus (e.g. the tips of the fronds). The parenchyma cells just inside the epidermis are suitable ones for measurement.

References

Isaac, W. E. (1933). Some observations on the drought resistance of *Pelvetia caniculata*. *Ann. Bot.*, 47, 343–348.

Isaac, W. E. (1935). A preliminary study of the water loss of *Laminaria digitata* during inter-tidal exposure. *Ann. Bot.*, 49, 109–117.

Zaneveld, J. S. (1937). The littoral zonation of some Fucaceae in relation to desiccation. *J. Ecol.*, 25, 431–468.

Investigation 2: Resistance to desiccation in animals

Most inter-tidal animals are also capable of tolerating and resisting a certain amount of desiccation and, like algae, their ability to do so has been correlated with position on the shore (Table 2; Lewis, 1964). This is well illustrated within the single

TABLE 2

The effects of desiccation on *Nucella* and *Littorina* spp. after 7 days at 18 ° C. (After Lewis.)

Species	Total water loss as % original wt.	Average % water loss/day	% Mortality	Position on the shore
Nucella lapillus	37·2	5·31	100	Middle shore
Littorina obtusata	56·5	8·35	80	Middle and lower shore
L. littorea	37·5	5·35	70	Middle shore
L. saxatalis	39·7	5·60	8–17	Middle and upper shore
L. neritoides	26·0	3·71	—	Splash zone

genus of *Littorina*. *L. neritoides*, which is common in the splash zone of many rocky shores, survives well in dry air and, to a lesser extent, so does *L. saxatalis*, which is also found high up the beach. However, the flat periwinkle, *L. obtusata*, and the edible one, *L. littorea*, which are both often abundant at about mid-tide level, lose water more rapidly in air and die more readily than their high-shore relatives.

Water loss by these periwinkles and, for that matter by many other gastropods, is far less severe than that of algae. These

differences are largely due to the ability of snail-like m
withdraw into their impervious shells in dry air and to cl
with their opercula. But not all animals lose water as sl
periwinkles. Many, such as starfish and crabs, have
exposed surfaces from which water is evaporated and lose it
more rapidly.

(a) Water loss in air and tolerance of desiccation

Most inter-tidal animals are suitable for these experiments. It is
interesting to compare the rate of water loss in a variety of
organisms, such as crabs, gastropods and starfish, and to con-
sider the results in the light of mechanisms of water retention
(e.g. impervious shells, etc.). It is also interesting to collect
animals from different tidal levels and to investigate the correlation between the resistance to desiccation and position on the
seashore.

As in the previous investigation, loss of weight can be used as
an index of the water that is lost by evaporation. Animals should
be weighed at the start of the experiment and thereafter at about
hourly intervals. However, if the investigation is restricted to
gastropods water loss is so slow that daily weighings are probably
sufficient. Five or more individuals should be used for each
species. Fortunately, weighing many inter-tidal animals is
scarcely more difficult than weighing pieces of algae. Many of
them can be kept and weighed in open beakers; most molluscs,
such as periwinkles and limpets, are inactive in dry air and do
not attempt to climb out of them, while more active animals,
such as crabs, cannot scale the sides of a beaker. Before the start of
the experiment, but not of course at each weighing, individuals
can be dabbed with absorbent paper in order to remove excess
moisture from them.

When comparing the loss of weight in different animals it is
worth remembering that the ability to tolerate desiccation (i.e.
to survive for long periods in spite of it) is also important. It is
valuable, therefore, to assess the ability of each species to survive
long periods of exposure to air. Clearly, this can be done as part

of a weighing experiment by determining the mortality rate for each species at the end of it. However, depending on the species used, it may be necessary to prolong the experiment because some inter-tidal organisms can live for several days in air (Table 2). It is usually obvious when active animals, such as crabs, are dead. Other animals can be tested for signs of life by placing them in sea-water and, thereby, giving them the opportunity to recover; gastropods soon come out of their shells, starfish move their tube feet and so on (see also page **15**).

(b) *The importance of the shell in water retention in limpets*
The thick impervious shells of limpets and their habit of clamping firmly down on rocks are undoubtedly important in preventing severe water loss from their bodies. The value of the shell and this clamping habit can be demonstrated by comparing the loss of weight in limpets that are allowed to clamp down in the normal manner in the beakers or petri dishes, in which they are kept for weighing, with others that are upturned so that their soft parts are exposed to the air. Limpets that are upturned cannot right themselves. Weighings should be made at intervals of about 4–8 hours and the mortality rate assessed in the two groups after one or two days.

The species *Patella vulgata* is abundant on many rocky shores and occurs throughout a large tidal range, so that some individuals are exposed to the air for much longer than others. Preliminary experiments by the authors suggest that within this species there may be a correlation between the rate of water loss and position on the seashore; individuals from high-up the shore tend to lose water less rapidly than those from nearer low water. This can be investigated by collecting groups of about ten or more limpets from the extremes of their tidal range and comparing the rate of loss of water in the manner described above. Limpets from high up the shore also have thicker shells than those from nearer low water (see Investigation 4), and there may also be a direct correlation between shell thickness and the rate of loss of water.

Reference

Lewis, J. R. (1964). *The Ecology of Rocky Shores* (Chapter 14). English Universities Press Ltd.

Investigation 3: Temperature tolerance in marine animals

The temperature of the air fluctuates much more than that of the sea, so that organisms that are exposed to the air for long periods (i.e. those living high up the shore) are likely to be subjected to greater extremes in temperature than organisms living lower down the shore. In exceptionally hot or cold weather the temperatures experienced on the shore are likely to be near lethal ones and, not surprisingly therefore, research workers have established a relationship between the ability of inter-tidal organisms to survive extremes in temperature and the position normally occupied on the shore. In general, organisms from high up the shore can tolerate a greater range of temperature than those living nearer low water (Table 3; Evans, 1948). There are certainly exceptions to this rule, but these can sometimes be accounted for when the precise nature of the habitat occupied is taken into account. For example, the limpet, *Patella vulgata,* has a lower temperature tolerance than its position on the seashore would suggest, but above mid-tide level it usually occurs in shaded and damp places, so that it is unlikely to experience the extremes of temperature that occur in more exposed places in the same zone.

Temperature tolerance can be investigated by determining the range of temperatures over which an organism can survive. The extremes (i.e. the thermal death points) can be worked out for both high and low temperatures, although, presumably because it is more convenient to study the effect of high temperatures, most investigators have confined themselves to a study of these. The investigations suggested here are also concerned only with high temperatures, although if suitable facilities are available an interested student could study survival at low temperatures.

It is rather ambitious to attempt a study of a large number of different species, and unless ample time is available, it is probably

TABLE 3

Temperature tolerance of some littoral animals. (Based on data from Southward, and Evans.)

Species	Lethal point (50% mortality) in ° C.	Normal position on shore
Periwinkles:		
Littorina neritoides	46·3	Splash zone
L. littorea	46·0	Middle shore
L. saxatalis	45·0	Middle and upper shore
L. obtusata	44·3	Middle and lower shore
Limpets:		
Patella depressa	43·3	Middle shore
P. vulgata	42·8	Middle and upper shore
P. aspera	41·7	Lower shore
Barnacles:		
Chthalamus stellatus	52·5	Upper and middle shore
Elminius modestus	48·3	High middle shore
Balanus perforatus	45·5	Middle to lower shore
B. balanoides	44·3	Lower shore

better to restrict the investigation to two or three species that are representative of widely separated zones on the shore.

The lethal temperatures of animals are usually determined by raising the temperature slowly until the death point is reached. This is normally done when they are submerged in water, because other factors, such as the loss of water by evaporation, may contribute to death in air. Ideally a thermostatically controlled water-bath should be used in these experiments, but other means of slowly heating sea-water can be devised. It is probably better to determine the lethal temperature of one species at a time. Several beakers, say six, containing sea-water and the experimental organisms should be heated slowly in the water-bath, starting at a temperature that is well below the expected lethal temperature (see Table 3). In order that good comparative results are obtained, it is important to regulate the rate of temperature increase, because this can affect the ultimate death point; a suitable way is to raise the temperature at a rate of, say, 2° C. every 5 minutes. It is not usually possible to determine the death point by observation, because many animals, particularly gastropods, go into a heat coma some time before death occurs. It is necessary, therefore, to go to some lengths to ensure that death has occurred. When death is first suspected some of the organisms should be tested for signs of life by pricking their soft parts with a needle or seeker. If the animals are apparently dead one beaker should be removed at that temperature and another at each successive 1° rise. These beakers should be left at room temperature for about 12 hours so that the animals in them are given the opportunity to recover. Any that still do not show signs of life can be assumed to be dead. The mortality rate should be determined for each beaker and the lethal temperature, which is usually considered to be the temperature at which 50% of the population dies, assessed for the species.

References

Evans, G. R. (1948). The lethal temperatures of some common British littoral molluscs. *J. Anim. Ecol.*, 17, 165–173.

Southward, A. J. (1958). Note on the temperature tolerances of some inter-tidal animals in relation to environmental temperatures and geographical distribution. *J. mar. biol. Ass. U.K.*, 37, 49–66.

Investigation 4: The shape and thickness of shells of limpets (*Patella*) and their relation to position on the shore

It may be convenient to study the radula length of limpets (Investigation 11; page 36) at the same time as this investigation.

Limpets occur throughout most of the tidal range on many rocky shores. Orton (1928) has shown that the shells of those living high up the shore tend to be taller and thicker than the shells of those living nearer low water (Figure 4). Shell thickness is believed to be concerned with heat insulation and also with water conservation (see Investigation 3), and the high-level individuals presumably benefit from their thicker shells when they are exposed to air, because they lose less water and are better insulated than low-level limpets. Their habit of clamping firmly down on rocks when exposed to the air also reduces the rate of evaporation of water from their bodies. However, this habit

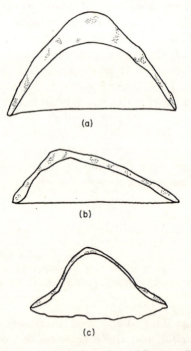

(a)

(b)

(c)

Figure 4. Shell shapes, in vertical section, of limpets, *Patella vulgata. a* is an individual from near high water level; *b* is a low-water individual; and *c* shows the ledging of a shell after transference from an exposed surface to a rock pool. (Drawn from photographs by Moore.)

has an indirect effect on growth of the shell. Limpets living high up the shore are exposed to air for most of their lives, so that their shells tend to be secreted when they are in a contracted state. As a result, they tend to become relatively tall and thin. On the other hand, the shells of limpets from near low water tend to be secreted when these animals are relaxed because they are covered with water for most of the time and, in this case, they are more flattened.

(a) Shell thickness

Limpets (*Patella vulgata*) can be collected in batches of, say, thirty from stations near mid-tide level, low water and high water. They can be removed from rocks by giving them a sharp blow or prized off with a stout knife.

The contents of the shells should be removed with a knife and the insides scraped clean. Thickness of the shells can be measured with a pair of callipers or a micrometer screw gauge, as long as it has fine points. Measurements should always be made at the same place. A suitable one is halfway between the tip of the spire and the anterior margin of the base.

(b) The shape of shells

There is no difficulty in measuring the height and length of shells, particularly if a pair of sliding callipers is available. The ratio of shell height/length should be determined for each group of limpets, because this itself gives an indication of the changes of shape in shells from different tidal levels. However, a more instructive method may be to measure the angles at the apexes of the shells. This can be done by placing each shell on paper and marking the tip of the spire and the most anterior and posterior margins of the shell with pencil marks on the paper. These marks should be joined to make the required angle, which can be measured with a protractor.

(c) Transference experiments

Moore (1934) showed that when limpets are transferred from one part of the shore to another there is a concomitant change in the growth of their shells. The shells of individuals from near low water, which are flattened, tend to become taller when they are transferred to a position higher up the shore. Conversely, high-water individuals have a more flattened growth after transference. This change in growth tends to leave a ledge around the shell and, as a result, was known as ledging by Moore (Figure 4).

These experiments can be repeated if a suitable shore can be visited at intervals. Groups of limpets of less than 25 mm in diameter should be collected from widely separated stations, preferably from near high water and from near low water. Individuals should be clearly marked with dabs of cellulose paint, using different colours for the two groups. Half the low-water individuals should be placed on suitable rocks near high water and the other half, as controls, returned to their original zone. Similarly, half the high-water limpets should be transferred to near low water and the others returned to the rocks near high water. About six months or a year later searches should be made for the marked limpets and shell shapes compared for the different groups.

References

Moore, H. B. (1934). The relation of shell growth to environment in *Patella vulgata*. *Proc. malac. Soc. Lond.*, 21, 217–222.

Orton, J. H. (1933). Studies on the relation between organisms and the environment. *Trans. Liverpool Biol. Soc.*, 46, 1.

Investigation 5: The shape and thickness of shells of dogwhelks, *Nucella (Thais) lapillus*

The conditions on sheltered and exposed shores are often so different that they have profound effects on the organisms living on them. For example, the channelled wrack, *Pelvetia caniculata*, is usually only found on exposed shores where there is enough spray to keep the splash zone moist with sea-water. On the other hand, large crabs, such as *Carcinus maenas, Cancer pagurus* and *Portunus puber*, are much more common in sheltered places than those that are subjected to severe wave action. There may also be differences in populations of the same species that live on exposed and sheltered shores. One example is provided by the dogwhelk, *Nucella lapillus* (sometimes known as *Purpurea lapillus* and, sometimes, *Thais lapillus*). The shells of dogwhelks living on shores that are exposed to a great deal of wave action tend to be thinner but have wider apertures than those of dogwhelks from more sheltered places (Figure 5). These differences are believed to be

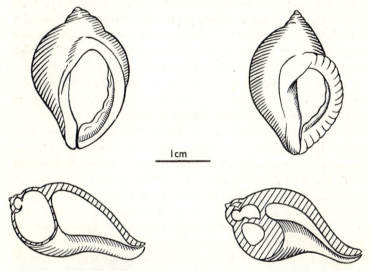

I cm

Figure 5. The differences between dogwhelks, *Nucella lapillus*, from an exposed shore (*left*) and a sheltered shore (*right*). (After Ebling, Kitching, Muntz and Taylor.)

related to the different conditions on exposed and sheltered shores. Dogwhelks living in exposed places must be able to resist the buffeting action of waves. To do this they must adhere to rocks firmly, and the large shell aperture, which indicates a large foot, reflects this ability. In sheltered places, where wave action is less important, a smaller foot is adequate and, concomitantly, the shell aperture tends to be smaller. However, large predators of dogwhelks, such as crabs (see above), often abound in such places, so that the thick shell can be seen as an adaptation to resist the crushing action of their claws (Kitching, Muntz and Ebling, 1966).

There is convincing experimental evidence to support these hypotheses. It was found that when crabs, *Portunus puber*, were kept in cages with dogwhelks from both sheltered and exposed places they crushed and ate far more of those from exposed places than from sheltered localities. However, sheltered-shore dogwhelks did not fare well when they were transferred to rocks exposed to severe wave action, presumably because their smaller feet had poor powers of adhesion and they were dislodged. In one experiment by Kitching, Muntz and Ebling, 100 individuals from a sheltered site and 100 from an exposed place were placed on an exposed rock, but whereas 90 of the dogwhelks from the exposed shore were still attached to it after 4 days, only 22 individuals from the sheltered shore still remained.

It is not always easy to find suitable exposed and sheltered shores that are close to one another, although there are undoubtedly such places (e.g. Lough Ine in Ireland, where Kitching *et al.* carried out their investigation). However, collections can be made from localities at some distance from one another (while on holiday?), preserved and examined at some later date. Suitable places for collecting dogwhelks occur, for example, in Cornwall; there are many good exposed coves on the north coast, which is open to the Atlantic, and many good sheltered ones on the south coast. Samples of about 30–50 dogwhelks should be made from, say, mid-tide level at each place. They can be preserved in 5% formalin or, alternatively, cleaned by boiling and removing the soft parts with a bent pin or seeker.

Measurements of the width and length of the shell aperture should be made with callipers. This presents no great difficulty, although it is necessary to decide whether or not to include the siphon groove in the measurement of aperture length. Shell thickness is more difficult, and it is necessary to break each one. Measurements must, of course, be taken consistently in the same place. This can be done by marking the intact shell in a suitable place (e.g. half-way between the tip of the spire and the end of the siphon groove) with a pencil and then breaking the shell carefully with a hammer. Providing reasonable care is taken, there should be no difficulty in dismantling the shell so that the marked piece can be found and measured with a micrometer gauge or callipers.

The colours of dogwhelk shells

There are also interesting differences in the colours of dogwhelk shells from different localities. In some places there is an enormous variety of colours, including black, brown, purple and orange ones, as well as some that are banded with one of these colours and white. In other places the entire population consists of white individuals. Moore (1936) has suggested that these colour differences can be related to differences in diet. Coloured whelks can usually be found where there is an abundance of mussels, which are a source of food, and Moore believes that the colour pigments are derived from these animals. On shores on which mussels are absent dogwhelks feed on barnacles, and because of the lack of suitable pigments in their diet are white (Plate 1). Moore was able to demonstrate the importance of diet experimentally: he found that changes in food could lead to a change in the colour of any new shell formed. However, the banding on most naturally occurring shells is not the result of changes in diet because the coloured bands are at right angles to the lines of growth. In fact, the situation is probably complex and might well repay further investigation. It seems likely that pigments derived from the diet are necessary for the full expression of certain genotypes.

References

Ebling, F. J., Kitching, J. A., Muntz, L., and Taylor, C. M. (1964). The ecology of Lough Ine. XIII. Experimental observations of the destruction of *Mytilus edulis* and *Nucella lapillus* by crabs. *J. Anim. Ecol.*, 33, 73–82.

Kitching, J. A., Muntz, L. and Ebling, F. J. (1966). The ecology of Lough Ine. XV. The ecological significance of shell and body forms in *Nucella*. *J. Anim. Ecol.*, 35, 113–126.

Moore, H. B. (1936). The biology of *Purpurea lapillus*. 1. Shell variation in relation to environment. *J. mar. biol. Ass. U.K.*, 21, 61–89.

Investigation 6: The burrowing habit of polychaete worms

Most of the characteristic organisms of sandy and muddy shores, particularly worms and bivalve molluscs, live buried in the substratum and are only found by digging and sieving. Some of them live in permanent or semi-permanent burrows, but others, such as the polychaete *Nephtys*, wander through the substratum.

There are probably several advantages in the burrowing habit. One is that, at depths in the sand, there are not the same fluctuations in temperature as at the surface. Similarly, the possibility of desiccation is greatly reduced. Burrowing animals are also afforded some degree of protection from surface-living predators, such as crabs, fishes and birds.

Burrowing can be investigated in a number of polychaete worms. *Nephtys* and the lugworm, *Arenicola*, are suggested here as suitable subjects, although others, such as ragworms, *Nereis*, can be used in similar experiments.

(a) Burrowing in Nephtys

Under suitable conditions *Nephtys* buries itself in sand very efficiently and burrows rapidly through it. The initial penetration of the substratum and subsequent progress through it are two distinct processes, involving entirely different types of activity.

When covered by water and provided with suitable substratum *Nephtys* penetrates it by one of two methods. While lying on the substratum, it may perform undulating movements which do not produce any appreciable forward movement but agitate the sand and sweep it on to the dorsal surface of the worm until it is buried. Alternatively, the worm may approach the sand obliquely, and when the prostomium (first segment) is in contact with it execute fast swimming movements. The agitations of the sand combined with forward thrust result in the first 15–20 segments being quickly buried.

Once the anterior end of the worm has penetrated the sand, burrowing through it involves use of the proboscis, which is everted punching holes in the substratum. In order to do this, the worm must be anchored, and this is done by the parapodia,

Plate 1. Dogwhelks, *Nucella lapillus*. The lower group was collected from mussel beds and the upper group from rocks on which barnacles were the only source of food. See the text for explanation.

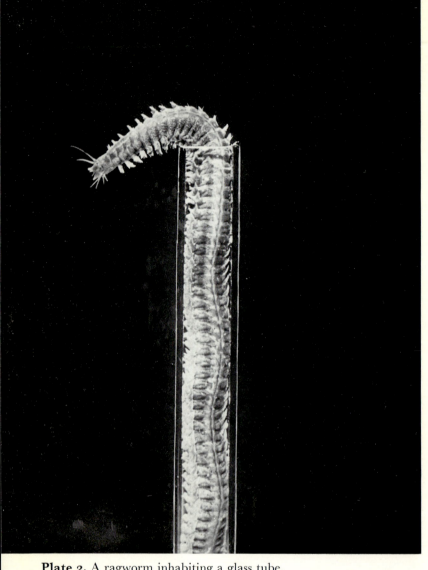

Plate 2. A ragworm inhabiting a glass tube.

which are held out at right angles to the body, preventing backward movement. Each time the proboscis is withdrawn the worm crawls into the hole made by it. The process is, of course, repeated cyclically as the animal progresses through the sand.

Initial penetration of the substratum can be observed if worms are placed on sand from their normal habitat in a small aquarium or large glass beaker; the sand should be covered with sea-water to a depth of 3–4 in. Burrowing can also be observed if worms can be persuaded to penetrate the sand near to the glass sides of the container so that they burrow down the side.

Nephtys usually has difficulty in burrowing into coarse sand if this is provided instead of the muddy sand from the animal's natural habitat. Apparently this is because the sand grains lack cohesion and the worm is unable to achieve the necessary anchorage with its parapodia.

(b) Burrowing in Arenicola

If a lugworm, *Arenicola*, is placed on wet sand or mud it normally burrows into it. Like *Nephtys*, the proboscis is important. It is extruded and withdrawn, punching holes in the substratum while peristaltic-like wave movements travelling along the body propel the animal forwards.

It was once thought that lugworms swallow sand as they burrow. This hypothesis can be tested by weighing individuals before and after they have burrowed into muddy sand. The worms can be weighed in beakers, and should be dabbed with absorbent paper before weighing to remove excess moisture. An alternative experiment is to keep some lugworms in clean sea-water for about a week so that they are starved and their gut becomes completely empty. Then they are allowed to burrow before being killed and dissected for signs of sand in the gut.

References

Clark, R. B. and Clark, M. E. (1960). The ligamentary system and the segmental musculature of *Nephtys*. *Quart. J. micros. Sci.*, 101, 149–176.

Wells, G. P. (1944). Mechanisms of burrowing in *Arenicola marina* L. *Nature*, 154, 396.

c

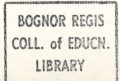

Investigation 7: The tubiculous behaviour of ragworms, *Nereis*

The behaviour of many burrowing worms is highly adapted to their mode of life. Some of them, such as the ragworm *Nereis*, will live in glass tubes in the laboratory and within them carry out apparently normal behaviour (Plate 2). This can be observed with little difficulty.

N. diversicolor, which is often abundant in estuarine mud flats and can be obtained in large numbers by digging, is suitable for these investigations as well as the other common species, *N. virens*, *N. pelagica* and *Perinereis cultrifera*. Worms left in a sea-water container for a few hours will usually enter and inhabit glass tubes provided for them. The tubes must, of course, be of appropriate sizes; ideally they should be slightly larger in diameter than the worms, but as the worms collected will vary in size, it is advisable to provide suitable lengths of tubing, say 10–15 cm, in a variety of sizes. Any worms that do not enter tubes of their own accord can be persuaded to crawl into them out of water; then when they are placed back in the water they will usually remain in them.

(a) *Irrigating behaviour and feeding*
Ragworms irrigate their tubes with undulating movements of the body which draw a current of water from head to tail. Wells and Dales (1951) have shown that, when worms are left undisturbed, bursts of irrigating activity occur at regular intervals. Periodically, they also reverse in their tubes and pump water in the opposite direction. Wells and Dales used a special recording apparatus, but continuous recordings can also be made by carefully observing a worm's behaviour for an hour or more. Longer recordings can, of course, be made by several observers relieving one another at intervals.

Sometimes worms can be persuaded to feed on small pieces of mussel, or other animal matter, which is placed near the end of the tube. They react by crawling towards the food, everting the pharynx and seizing it with the jaws. Often they withdraw back into the tube before devouring the food. However, these feeding experiments are only likely to be successful if the ragworms are not disturbed by other sources of stimulation. Furthermore, newly collected worms will not usually feed in this way; worms that have inhabited tubes for a week or more are far more likely to feed.

(b) *The withdrawal reflex and habituation*

Like many other polychaete worms, ragworms have a rapid withdrawal reflex that is elicited by almost any sudden stimulus, such as a shadow or tactile stimulus. Under natural conditions it is an escape response against predators and enables worms extended from their burrows to withdraw into them whenever danger threatens. Worms inhabiting glass tubes in the laboratory will also withdraw in response to sudden stimulation, even when they are not extended from their tubes. A blunt seeker can be inserted into a tube and used to stimulate a worm tactually, and shadows can be caused by passing a solid object, such as a book, between the worms and a bench lamp above them. (Worms usually need to be light adapted for about a quarter of an hour before they will react to shadows.)

If the same stimulus is presented on a number of occasions in fairly quick succession the worms become habituated to it. That is to say, they learn not to react to it. Sometimes rapid withdrawals occur to only the first one or two in a series of stimulations. The rapid responses are then replaced by slower, less obvious contractions, which themselves eventually disappear so that the worms do not respond at all.

Habituation, which is usually considered to be the simplest form of learning, can be investigated to either of the stimuli mentioned above or others, such as mechanical shocks caused by dropping a heavy object on to the bench near the worms. (See the paper by Clark (1960b), who compared the rates of habituation in worms subjected to a variety of different stimuli.) The best results are obtained if the worms can be kept in a room where they are relatively free from external stimulation (or light adapted in the case of shadows) for about an hour before the experiments. They should be stimulated at intervals of about a minute. When habituation has occurred, it is necessary to ascertain that the failure to respond has not resulted from muscular fatigue. This can be done by showing that the worms are still capable of responding to a stimulus other than the one to which they have habituated. Worms that have habituated to, say, shadows should be capable of responding to tactile stimuli, whereas fatigued worms would be incapable of responding to either stimulus.

References

Clark, R. B. (1960a). Habituation of the polychaete *Nereis* to sudden stimuli. 1. General properties of the habituation process. *Anim. Behav.*, 8, 83–91.

Clark, R. B. (1960b). Habituation of the polychaete *Nereis* to sudden stimuli. 2. Biological significance of habituation. *Anim. Behav.*, 8, 92–103.

Wells, G. P. and Dales, R. P. (1951). Spontaneous activity patterns in animal behaviour: the irrigation of the burrow in the polychaetes *Chaetopterus variopedatus* and *Nereis diversicolor* O. F. Muller. *J. mar. biol. Ass. U.K.*, 29, 661–680.

Investigation 8: Maintenance of position on the shore by the periwinkle, *Littorina littorea*

Relatively little is known about the ways in which most motile organisms maintain their positions, and therefore zonation, on the seashore. It can be shown that most gastropod molluscs confine their movements to their particular zones by marking individuals with dabs of quick-drying paint and recording their positions from time to time. Some organisms also have the ability to return to their correct zones if they are moved from them. This can be investigated by moving marked individuals to new positions on the shore and then recording their distribution some time later. The number of marked individuals that are recaptured is often small, so that large numbers should be used in these experiments.

By recording the movements of marked individuals, Newell

(1958) has shown that the edible periwinkle, *Littorina littorea*, remains in more or less the same position on the shore for many weeks. This is not the result of inactivity, because these animals make regular feeding excursions. However, on many sandy and muddy beaches tracks are left by the periwinkles, and Newell found that these were usually U-shaped. Apparently the periwinkles set off in one direction and then turn round, roughly counter-marching the original tracks so that they end up more or less at the setting-off place (Figure 6). Newell showed that they do this by orientating to the sun's rays by simple responses known as taxes. These are responses in which the animal moves in a fixed direction to the source of stimulation; usually movement is either directly towards or away from the source. He found that the majority of tracks were either towards or away from the sun, which itself suggests that they are orientating to it. That they were doing so was confirmed by shielding periwinkles that were on feeding excursions from the sun and then reflecting the light on to them from the opposite side with a mirror. They responded by turning round and reversing their direction of movement.

On some beaches, particularly those near the mouths of estuaries where there is a certain amount of silt, the tracks of periwinkles are not at all difficult to find. The directions of the tracks can be determined with the aid of a compass. Similarly, Newell's experiments of shading periwinkles on feeding excursions and reflecting the light on to them with a mirror can be repeated.

References

Newell, G. E. (1958a). The behaviour of *Littorina littorea* (L.) under natural conditions and its relation to position on the shore. *J. mar. biol. Ass. U.K.*, 37, 229–240.

Newell, G. E. (1958b). An experimental analysis of the behaviour of *Littorina littorea* (L.) under natural conditions and in the laboratory. *J. mar. biol. Ass. U.K.*, 37, 241–266.

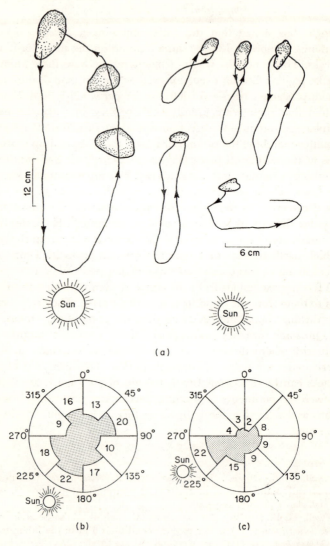

Figure 6. *a*: the tracks made by periwinkles, *Littorina littorea,* crawling on sand in bright sunshine. *b* and *c*: polar diagrams to show the direction of movement and numbers of periwinkles moving along various tracks. (After Newell.)

2

Feeding

In terms of the numbers of living organisms, the shallow littoral waters are undoubtedly the most productive of the whole sea. Plants, including many large algae, flourish in inshore regions where they can attach themselves to the substratum and still receive enough light to carry out photosynthesis. In waters that are deeper than about 100 m the penetration of light is poor and bottom-living plants cannot survive. Near the shore plant-life is also encouraged by the constant supply of nutrients which are washed into the sea from the adjoining land.

There are almost always complex feeding inter-relationships between organisms living on the shore and, if desired, food chains and webs can be worked out. The feeding habits of many of the common animals are well known but, for those that are not, it is interesting to dissect out and examine the gut contents. Identification of partly digested remains is sometimes difficult, but it is often possible to recognise crustacean limbs, chaetae of worms and so on. Indeed, given appropriate keys, identification to species can sometimes be made from such remains.

As might be expected from the abundance of plants on the shore, herbivorous animals are also well represented. However, surprisingly enough, some of the larger fixed algae do not appear to be fully utilised as a source of food. Some animals, such as the blue-rayed limpet, *Patina pellucida*, which leaves feeding tracks on the fronds and stipe of the oar-weed, *Laminaria*, undoubtedly feed on them, but most algae seem to suffer only trivial interference from animals. Small algae and plankton, which is also rich in inshore waters, are more fully exploited. There are, for instance, several gastropods, such as the limpets, *Patella* spp. and periwinkles, which rasp algae from rock surfaces and may clear large areas of rock. There are also many filter feeders, including bivalve molluscs, sea squirts and barnacles, which remove organic particles and micro-organisms from sea-water by a variety of filtration methods, involving the use of cilia, mucous nets or specially modified appendages. Another source of food is detritus, which is often exploited by burrowing animals. A common example is the lugworm, *Arenicola marina*, which feeds on the muddy sand (containing detritus) in its burrow. Because of the high undigestible content of this diet, large quantities of faeces are produced; these are deposited as casts near the exits of the burrows.

Carnivores, too, are abundant on the shore, and the feeding habits of some of them, such as starfish, crabs and fishes, can be observed if they are kept in sea-water aquaria. Others leave evidence of their feeding activities on the shore. For example, dogwhelks, *Nucella lapillus*, bore holes with their specialised radulas in the shells of mussels on which they prey, and empty shells with these neat little holes in them are often found on the shore.

Investigation 9: Filter feeding in mussels, *Mytilus edulis*

Mussels, which filter particles of organic matter from sea-water by means of enormously enlarged gills, are particularly common on many rocky shores and also near the mouths of estuaries. The filtering action of the gills can be investigated in opened (but living) mussels by pipetting suspensions of coloured particles (e.g. carmine) on to them.

A current of water is drawn into the partly opened shell through one opening, the inhalant siphon, and out through another, the exhalant siphon (Figure 7). Each gill consists of a series of ciliated filaments. The arrangement of the cilia, of which there are three types, each with a different function, is shown in Figure 7d. The incurrent of water is drawn into the shell and through the gills by the powerful lateral cilia. Particles in this

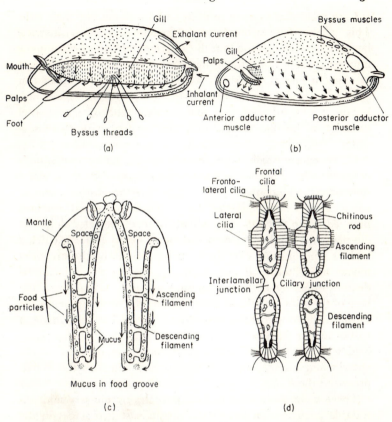

Figure 7. Filter feeding in the mussel, *Mytilus edulis. a* : food currents in an opened mussel. *b* : rejection currents in a mussel whose gills have been removed. *c* : a vertical section through the gills of one side. *d* : a horizontal section through two adjacent filaments. (After Borrodaile, Potts, Eastham and Saunders.)

current which come into contact with the gills are deflected by the latero-frontal cilia on to the frontal cilia, which beat downwards, carrying a stream of them to the base of the gill. Here they are bound in a continuous stream of mucous which passes along the food groove (Figure 7*c*) to the mouth.

Care is needed to open mussels without damaging their gills. This can be done successfully if a strong pen-knife is inserted between the two shells so that the anterior and posterior adductor muscles, which hold them together (Figure 7*b*), can be cut as close to one of the shells as possible. The two halves can then be prized open to display the gills. (Scalpel blades are usually too brittle for this and break.) The opened mussel should then be placed in a dish and covered with clean sea-water.

Unless exceptionally large mussels are available, observations of the filtering processes will be greatly enhanced by the use of a stereoscopic (binocular) microscope. Suspensions of carmine particles in sea-water should be prepared and small amounts pipetted on to the gills. Movements of these particles along the gill filaments can be observed without difficulty (other particles, such as sand or mud, can also be tried). Similarly, the stream of mucus along the food groove at the base of the gill can be demonstrated by cutting a small V-shaped portion from the base of the gill with scissors and rejoining the broken mucus cord so that it flows across the gap. When the cut has been made, a ball of mucus forms at the posterior cut surface. This should be held with forceps and pulled gently across to the base of the anterior cut surface, where it will usually join with the mucus produced there.

There is some selection of food particles by mussels, and unsuitable particles can be rejected at at least two stages in the filtering process. First, excessively large particles are rejected because the frontal cilia cannot carry them. They fall from the gills on to the mantle cavity wall, where rejection ciliary currents direct them to the exhalant siphon. Second, the labial palps can deflect the mucus stream away from the mouth if the food is unsuitable. Carmine is usually rejected in this way; balls of mucus, coloured red with carmine, form on the palps before eventually falling on to the mantle wall, where they may be carried away in the rejection currents.

Small pieces of gill can be examined microscopically if

they are mounted in sea-water. If the preparation is flooded with carmine the ciliary action in moving the particles can be observed.

Investigation 10: The feeding behaviour of sea anemones

A sea anemone captures living prey and pieces of dead animal matter with its tentacles. When food touches a tentacle, nematocysts, which penetrate and hold the food, are fired into it. The tentacles then bend over and push the food into the mouth.

It is interesting that such well co-ordinated and elaborate behaviour can be achieved in an animal with such a simple nervous arrangement, for, like other coelenterates, the nervous system of an anemone consists of diffuse nerve net with no centralised control. The nematocysts are also interesting because they provide one of the few examples of what are known as independent effectors. These are cells that can both detect changes in the environment (stimuli) and react to them.

The experiments suggested in this section are concerned with: (*a*) the chemical sense of anemones; (*b*) the reactions of isolated tentacles; and (*c*) with nematocysts. Experiments on the chemical sense can be performed with the beadlet anemone, *Actinia equina*, on the seashore (Plate 3). This animal is abundant on many rocky shores. It has the disadvantage that it can withdraw its tentacles and, indeed, it always does this out of water. Anemones in this condition are unsuitable for the experiments, but those in rock pools which are 'open' can be used. The snakelocks anemone, *Anemonia sulcata*, cannot withdraw its tentacles, but it is restricted in its distribution to the South and West coasts. However, if it can be obtained, it is well suited to both field and laboratory studies, where the additional experiments (*b*) and (*c*) can be carried out. Individuals kept in the laboratory should be allowed to establish themselves in well-aerated sea-water aquaria. If they can be persuaded to attach themselves to glass plates of about 4 in.^{2} they can be transferred to smaller containers for experimental purposes.

(a) The chemical sense of anemones

It can be shown that sea anemones recognise their food by a combination of chemical and mechanical stimulation. The response to a purely mechanical stimulus, such as that from a blunt seeker, is usually a localised one. The tentacles that are actually touched perform a few exploratory movements but then move away from the seeker. However, if a piece of dead animal matter (e.g. a piece of limpet or mussel) is placed on the tentacles of an anemone so that it provides both chemical and mechanical stimulation there is a more pronounced response. The tentacles that are already in contact hold the food, while nearby ones bend over until they, too, come into contact with it. The mouth then opens and the food is pushed into it.

The chemical sense of anemones can be investigated further by testing their reactions to small balls of cotton-wool soaked in various substances, offered to them with forceps. They usually react to cotton-wool soaked in sea-water in the same way as a blunt seeker, except that there is sometimes active rejection of the ball; the tentacles may bend away from it so that it is exposed and rolls off the oral disc. As might be expected, anemones react to balls soaked in animal juices as if they were food and ingest them. Other cotton-wool balls can be soaked in a variety of substances (e.g. saliva, sugar) and offered to anemones. Surprisingly, they will often ingest balls of cotton-wool soaked in saliva but reject those soaked in sucrose solution. The explanation seems to be that, in general, anemones react positively to stimuli that are likely to indicate food in their normal life. Thus mucus, which is, of course, a constituent of saliva, is likely to be associated with food under natural conditions, because many marine animals are coated in it. Sucrose, on the other hand, is unlikely to be encountered naturally.

If these experiments are to be carried out on the seashore balls of cotton-wool soaked in various substances can be prepared beforehand. Each ball should be offered to a separate anemone and the responses recorded. The following scale may be useful, although additional notes should always be made:

++++ more than 60% of the anemone's tentacles react to
      the ball;
 +++ 40–60% react to it;
  ++ 20–40% react to it;
   + less than 20% react to it;
   — no response;
   = rejection of the ball.

Large numbers of anemones cannot, of course, be used in
laboratory experiments, and so that individuals can be used
several times, it is important to prevent them from actually
ingesting the balls. (If they do, they become satiated and unreactive.) This can be done by removing the balls with forceps
before they are taken into the mouth.

(b) Responses of isolated tentacles in Anemonia sulcata

Because there is no centralised control of activities by a central
nervous system in coelenterates detached parts of an anemone,
such as an isolated piece of tentacle, still react as if they were
attached to the whole organism. This is in marked contrast to,
say, the limb of a vertebrate, which is unreactive after amputa-
tion from the central nervous system.

Tentacles can be removed from the snakelocks anemone with
scissors. Cut pieces should be transferred to dishes of clean sea-
water and left to recover for 5–10 minutes. Their reactions to
mechanical stimuli can then be tested by stimulating them with a
blunt seeker. Usually there is little response. However, if an
isolated tentacle is flooded with food juice (e.g. crushed mussel)
there are normally prolonged contractions. The tentacle may
even bend over towards the side on which the mouth was situ-
ated, as if to push away the food in that direction.

(c) Nematocysts

Like tentacles, the reactions of nematocysts can be fired by tactile
stimuli, but these are usually ineffective unless they are accompanied by appropriate chemical stimulation. The effective

chemicals are those that are normally associated with food. This can be demonstrated by touching the tentacles of anemones with glass coverslips which have been previously smeared with food juice or saliva and, as controls, clean coverslips. The food juice or saliva should be smeared on to the coverslip and then allowed to dry before the test is made. The coverslips should be examined microscopically after each test and the sensitivity of nematocysts estimated by determining the relative numbers of discharged nematocyst on clean coverslips and those smeared with food.

Reference

Pantin, C. F. A. (1952). Demonstrations of behaviour in the lower animals. *The Science Masters' Book*. Series III. Part III. *Biology*. pp. 144–185.

Investigation 11: The relationship between radula length and position on the shore in limpets

This investigation can be carried out in conjunction with Investigation 4 (see page **15**).

Several marine gastropods, including limpets, rasp algae from the surface of rocks with long strap-like radulas. Two investigators, Brian and Owen (1952), have shown that limpets living high up the shore tend to have longer radulas than those from nearer low water. The radula is gradually worn away with use, and Brian and Owen have suggested that the differences in length may be due to increased use in low-water individuals. Limpets are believed to feed almost exclusively when they are covered by water, so that those living low down the shore will have more opportunity to feed than those from higher up it. They may therefore use their radulas more, although it still needs to be established that they actually do so.

This investigation can be carried out with *Patella vulgata*, which is extremely common and occurs throughout most of the tidal range. In order to compare radula lengths meaningfully it is necessary to take into account the overall sizes of limpets.

Brian and Owen did this by determining the radula fraction, which is the ratio of the radula length to shell length.

Groups of about 25 limpets should be collected from stations at different tidal levels: say, near high tide, mid-tide and low water. The shell length should be measured from extreme anterior to extreme posterior ends for each individual. The radula can be removed by first cutting open the buccal mass (anteriorly and posteriorly from the mouth) with scissors and then pulling it out with blunt forceps. At first the radula may be difficult to find, but with a little practice this becomes an easy operation. Radulas, which are surprisingly long (approx. 2–7 cm), can be stretched out on a tile and measured with a ruler. It is preferable to remove the radulas from limpets that have been killed, although they can be removed from live ones. A simple and quick way of killing limpets is to plunge them into boiling water. However, each individual must be killed separately, because the shell often becomes detached from the rest of the animal after this treatment. If the two parts do become separated from one another the radula fraction cannot, of course, be worked out.

Reference

Brian, M. V. and Owen, G. (1952). The relation of radula fraction to environment in *Patella. J. Anim. Ecol.*, 21, 241–249.

Investigation 12: The effect of grazing by limpets on the surrounding flora

Limpets usually inhabit permanent rock scars, to which they return after feeding excursions, and their grazing activities often have a profound effect on the surrounding plants. In places such as the walls of harbours, where there is a thick felt of *Enteromorpha* or other filamentous algae, there may be areas around the limpets' scars that are completely clear of algae (Figure 8). This situation affords an opportunity to investigate the food requirements of limpets, and Moore (1938) has shown that the area grazed by a limpet (= the area denuded of algae)

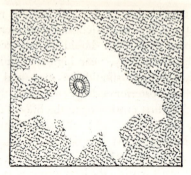

Figure 8. The browsing area of a limpet, *Patella vulgata*. The region still covered with algae is stippled. (After Moore.)

is more or less a linear function of its volume. The only exceptions to this rule seem to be the smallest animals, which, presumably because of their higher growth rates, take relatively more food than larger ones.

This problem can be investigated if a place can be found where the grazing areas of limpets are clearly visible. The volumes of limpets can be calculated if the length, breadth and height of each is measured and it is then assumed that each is a cone whose basal diameter is equal to the mean of the length and breadth. The feeding area must be estimated by taking measurements in the field, and then this can be plotted graphically against volume for each individual.   A simple way of estimating the areas is to take pieces of graph paper, marked in mm, into the field and to mark or cut these to the shapes of the limpets' grazed areas.

Reference

Moore, H. B. (1938). Algal production and food requirements of a limpet. *Proc. Malacol. Soc. London*, 23, 117–118.

Plate 3. Two beadlet anemones, *Actinia equina*.

Plate 4. The parasite *Peltogaster paguri* on the abdomen of a preserved hermit crab.

3

Parasitism and commensalism

There are many interesting inter-relationships between animals living on the seashore as well as in the sea itself. Many of these fall into the category of parasitism, which is sometimes defined as the association between two organisms, one of which (the parasite) obtains its food from the other (the host) without normally killing it. Some parasites of marine organisms are particularly interesting because they provide superb examples of the extreme adaptations to this highly specialised way of life. Several crustacean parasites have become so highly adapted that in external appearance they bear little resemblance to free-living animals in the same class, and only with a knowledge of their life histories has it been possible to classify them satisfactorily. For example, in the adult stage, *Sacculina carcini*, which is a common parasite of the green shore crab, consists of little more than an external sac attached to the crab's abdomen, from which tubes ramify throughout the host's body. It is quite unlike typical cirripedes (barnacles), and yet it is almost certainly a member of this order, because it goes through characteristic nauplius and cypris larval stages during its development (Figure 9).

There are also many commensal relationships on the seashore. These are usually considered to be associations from which one or both partners benefit but which are not obligatory, that is to say, the two organisms can live satisfactorily on their own. Hermit crabs, in particular, are well known for their commensal relationships with other animals.

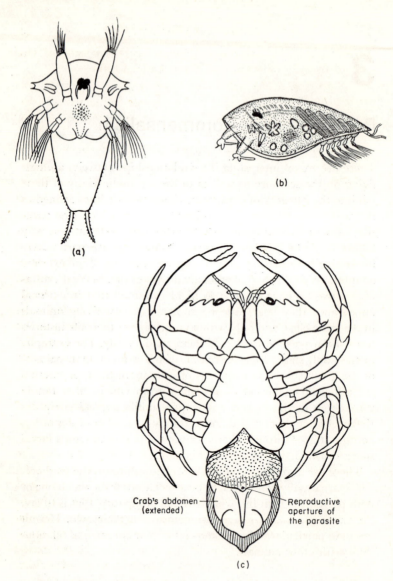

Figure 9. Stages in the life history of the parasite *Sacculina carcinii*. *a*: nauplius larva. *b*: cypris larva. *c*: an adult parasite attached to the underside of a crab's abdomen. (Not to scale.)

Investigation 13: Copepod (Crustacean) parasites of the cod, *Gadus callarius*, and other gadoid fishes

Although most of the well-known copepods, such as *Cyclops* and *Calanus*, are free-living animals, the group (sub-class) also includes some remarkable parasites. Linnaeus included one of them, *Lernaeocera*, in his group 'Vermes', but, like *Sacculina*, it gives away its true identity during its life history, which includes typical copepod stages (Figure 11).

Several common copepod parasites infect fishes of the family Gadoidea, which includes the cod, the haddock, the pollack and the whiting. Infestations of parasites on their hosts, particularly the cod, are frequently high, so that examinations of even relatively small numbers of fish are usually rewarding. It is often possible to examine recently caught fish at the fish markets of coastal towns; the authors, for example, have done this at Tynemouth and Whitby. Naturally it is necessary to ask permission to search fishes for parasites, but our requests have always been well received. Alternatively, discarded fish heads can be examined, because most of the parasites occupy the head and gill regions. This can be done at the fish market or, if it is possible to arrange for a fishmonger to supply the heads, in the laboratory.

Descriptions of copepod parasites are not given in textbooks, so that the three most commonly occurring ones are described below. Others can be identified from Scott and Scott (1913).

Adult *Caligus rapax* (Figure 10), which are about 5–10 mm in length, are usually found in the mouth of the host, but they are capable of movement and can occur almost anywhere on the surface of the fish. They can also move from one fish to another and, on occasions, have been recorded in plankton samples. Attachment to the host is by means of sucking discs at the anterior margin of the carapace, by clawed second antennae and large clawed maxillipeds. Mature females can be recognised by their two long egg tubes. Each egg hatches into a typical, free-swimming nauplius larvae. After a series of moults this develops into the so-called chalimus larva, which is the host-infective stage. It becomes attached to the gills of a suitable fish by a long fila-

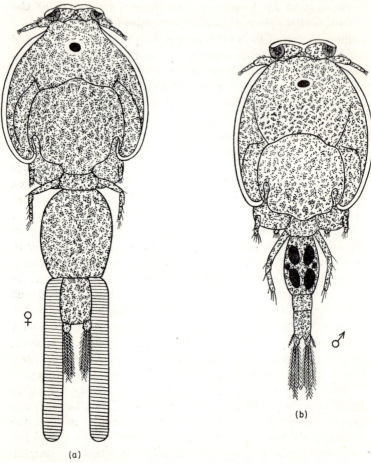

Figure 10. Adult male and female *Caligus rapax*. (Both ×13.)

ment secreted by glands in the head. There is a further series of moults before the adult stage is reached.

Lernaeocera is a large (up to about 4 cm), worm-like, blood-feeding parasite, which often projects from the gill chamber of its host. *L. branchialis* is parasitic on the cod, and *L. obtusata*, a similar species which has only recently been separated from *branchialis*, parasitises the haddock. The twisted body of the adult is attached to a gill arch by three deeply buried, branching processes, which absorb nourishment. The life history is more complex than that of *Caligus*, and it includes an intermediate host, which may be any one of a number of species of flatfish (Figure 11). The nauplius larvae attach themselves to the gills of the intermediate host, where they develop through a series of stages into sexually mature copepodid larvae. These mate precociously on the flatfish. After copulation the males die, but the females

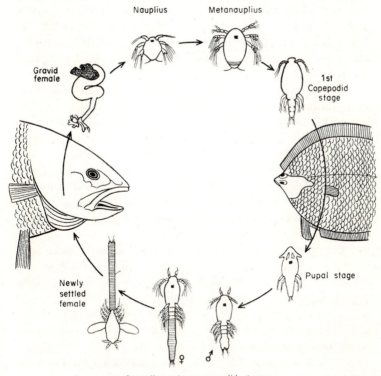

Figure 11. The life history of *Lernaeocera branchialis*. (After Cameron.)

move off in search of the definitive host, the gadoid fish. They attach themselves to the gills and mature into gravid females.

Clavella uncinata belongs to the same family (Lernaeidae) as *Lernaeocera*, and in the adult stage it also has a degenerate body structure, consisting of little more than a digestive tract and reproductive system (Figure 12). There is, however, only one host in the life history. The earlier larvae stages are suppressed and copepodid larvae hatch from the eggs. These seek suitable hosts, and attach themselves by puncturing the skin with their jaws and inserting the long head tube; at a later stage specially modified maxillae form the attachment. However, the males eventually free themselves and crawl over the hosts in search of females, with which they mate. They die after copulating, while the fertilised females develop two large egg-sacs. Adult females are about 6–8 mm in length, excluding the egg sacs; males are only about 1 mm long.

Infestations (i.e. the number of parasites per host) vary from one parasite to another and, of course, from one host to another. On occasions as many as 50 or 60 *Caligus* may be found on a medium-sized cod, and similarly infestations of 20 or more *Clavella* are not uncommon. However, *Lernaeorcera*, which probably has a more serious effect on its host, is usually found living alone. Spronston and Hartley (1954) record some multiple infestations of this parasite on haddock and pollack, but never more than 4 on the same fish. These authors suggest that multiple infestation by parasites provides evidence of the host's inability to build up an immunity and thereby resist further attacks.

Interesting data can be obtained by comparing the degree of infestation on hosts of different sizes. For example, Poulsen (1939), working on the infestation of *Clavella* on the cod, found that while the youngest fish (below 11 cm) were free of parasites, the highest rate of infection (38%) was on one-year-old specimens; thereafter there was a fall-off in the number of parasites with age. This suggests that there is a crucial stage, at about the end of the first year, at which infection occurs.

Spronston and Hartley (1954) have shown that, in the case of

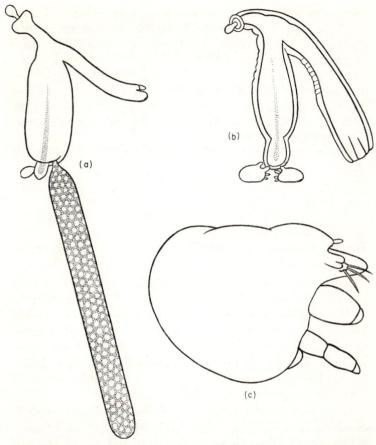

Figure 12. The parasite *Clavella uncinata*. *a*: a mature female. *b*: a juvenile female with two males attached to its abdomen. *c*: a mature male. (*a* and *b* both ×7; *c* ×45.)

Lernaeocera obtusata on haddock and pollack, infection occurs when the fish migrate to inshore waters, where the parasite's intermediate host, the flatfish, abounds. They found that young haddock and pollack, migrating inshore from deeper waters, were free of parasites, but that they became infected with them soon afterwards. *Lernaeocera branchialis* may infect cod at a similarly critical stage in the host's life, because the authors (S.M.E. & J.M.H.) have only found these parasites on large (more than 60 cm in length) fish.

References

Poulsen, E. K. (1939). Investigations upon the parasitic copepod *Clavella uncinata* (O. F. Müller) in Danish waters. *Vidensk. Medd. Fra Dansk. naturk. Foren.*, c11, 223–244.

Scott, T. and Scott, A. (1913). *The British Parasitic Copepoda*. Ray Society.

Spronston, N. G. and Hartley, P. H. T. (1954). The ecology of some parasitic copepods of gadoid and other fishes. *J. mar. biol. Ass. U.K.*, 25, 361–392.

Investigation 14: Animals associated with hermit crabs, *Eupagurus bernhardus*

Some of the most fascinating associations are those between hermit crabs and the animals that live either inside or outside the shells occupied by them. Some of the associates are actually attached to the crabs and are undoubtedly parasitic, but others are commensal, although in many cases there is still speculation about the benefits that the partners derive from their association. Original research into some of these relationships might be extremely rewarding.

Large hermit crabs, *Eupagurus bernhardus*, inhabiting the shells of whelks (*Buccinum*) are most suitable for study. Smaller crabs, which are often found in periwinkle shells and may be common inter-tidally, are less suitable, because there are usually fewer animals associated with these specimens. The larger crabs can be obtained by dredging the sub-littoral zone. If this cannot be arranged, preserved ones in their shells can be purchased from marine biological stations or from various biological suppliers.

The shells inhabited by crabs should be cracked and carefully dismantled with pliers. Some books suggest that live crabs can be persuaded to leave their shells by holding a lighted match beneath the tip of the shell, but at best this method is only partially successful, and even if the crab leaves the shell, the commensal worm, *Nereis fucata*, which may also be living in it, does not.

One of the most interesting parasites of *Eupagurus* is the crustacean, *Peltogaster paguri*, which belongs to the same sub-class (Cirripeda) as barnacles and *Sacculina*. The adult *Peltogaster* is a curved oval sac, about 1 cm long, which is attached to the side of the crab's abdomen (Plate 4). The externa is bright red, and roots, which penetrate the host at the point of attachment and radiate throughout the abdomen, are a conspicuous green. *Peltogaster* may itself be parasitised by another crustacean parasite, the isopod, *Liriopsis pygmaea*, which is also sac-like in appearance.

Of the several commensals that share the shells occupied by hermit crabs, the best known is probably the ragworm, *Nereis fucata*, which grows to a maximum length of about 10 cm and inhabits the terminal whorls of the shell. Thorson (in Caullery, 1952) constructed an artificial glass shell and observed the behaviour of a crab and ragworm living in it in dim red light (most marine animals are relatively insensitive to this). Normally the worm remained at the end of the glass shell, but when the crab started to feed it moved to the entrance and fed on pieces of food missed by the crab. The presence of the worm is apparently tolerated by the crab, because there is no attempt to capture it. It seems likely, therefore, that the crab actually benefits from the association. This may be derived from the ragworm's habit of irrigating (see Investigation 7), which presumably draws a current of water through the shell. Infestations of *Eupagurus* shells with *N. fucata* seem to increase in warmer waters. In samples from Sweden, for example, only 6·6% of crabs were associated with worms, from the West coast of Scotland 23·6% of shells had commensals in them, from the Normandy coast 30% were infected and from Portel 50%.

Numerous organisms live on the outsides of the shells and presumably benefit from the movement of the crab. Barnacles,

Balanus balanoides and *B. perforatus*, and the worms, *Spirorbis* and *Pomatoceras*, which live in calcareous shells, frequently form permanent attachments to the shells, and so do various hydroids, such as *Laomedia flexuosa* and *Hydractinia echinata*. The large anemone, *Calliactis parasitica*, is also found on the shells of hermit crabs. The crabs actually plant these anemones on their shells and to do so require the anemone's co-operation. When one of them encounters a calliactis it performs an elaborate sequence of acts whereby the anemone is manipulated with the claws and rolled on to its shell. Anemones are also transferred to new shells when the crab changes its home. This relationship is not obligatory, and both *Calliactis* and *Eupagurus bernhardus* are often found living on their own. However, a similar association between another hermit crab, *Eupagurus pridieauxi*, which can be purchased from some marine biological stations (e.g. Scottish Marine Biological Association, Millport, Scotland), and the anemone *Adamsia palliata* is taken a stage further. This anemone cannot live without the association, and dies if it is removed from the crab, even if it is fed artificially. *E. pridieauxi* lives in smaller shells than *bernhardus*, but the anemone grows round what would otherwise be exposed parts of the abdomen, thereby protecting them (Plate 5). In fact, the base of the anemone, which is hard and rigid, is specially adapted for this purpose.

Reference
Caullery, M. (1952). *Parasitism and Symbiosis*. Sidgwick and Jackson Ltd., London.

Investigation 15: The effect of the parasite *Sacculina carcini* on male green shore crabs, *Carcinus maenas*

Crabs, *Carcinus maenas*, parasitised with *Sacculina carcini* are common in many localities, particularly on the south and west coasts. They can therefore be collected from the shore or, alternatively, purchased from biological suppliers or marine stations.

The adult *Sacculina* consists of a white or yellow fleshy sac, which is attached to the underside of the thorax but is partly covered by the crab's abdomen (Figure 9). Numerous branching roots, which absorb nourishment from the host, radiate through the crab's body from the point of attachment. The reproductive organs, which occupy most of the external sac, mature in the spring and the life history, like that of other cirripedes, includes both nauplius and cypris stages. It is the cypris stage that actually infects a new host.

The parasite has an interesting effect on male *Carcinus*. It destroys the androgenic glands, so that 'male' hormones are no longer released. As a result, parasitised males tend to become more like females in appearance. The usual way of distinguishing males from females is by the shapes of their abdomens. Those of females are rounded and consist of 7 clearly separated segments, whereas those of males tend to be pointed and the middle 3 segments are fused together so that there are apparently only 5 of them. In parasitised males the abdomen tends to be intermediate (Plate 6). The animal can still usually be recognised as a male, but the abdomen is broader and more rounded than normal. Furthermore, there is a tendency for the fused segments to subdivide so that all 7 segments become clearly visible.

Reference
Caullery, M. (1952). *Parasitism and Symbiosis*. Sidgwick and Jackson Ltd., London.

4

Microhabitats: rock pools, crevices and animals associated with seaweeds

The substratum on sandy and muddy shores usually provides a fairly uniform habitat in which there may be gradual changes in the population from one part to another (e.g. from high to low water), but there are not normally sudden changes. However, on rocky shores, particularly where the substratum is broken and uneven, there may be a variety of microhabitats, such as rock pools and the crevices in rocks. Conditions in such places are usually quite different from those on adjacent parts of the shore, and as a result, they tend to support their own characteristic populations. Seaweeds, too, provide cover for many organisms that are unable to survive in exposed places, and they may harbour a number of animals that are not found elsewhere on the shore. These microhabitats make interesting subjects for investigation, particularly for small groups of students working on their own, and because they have so far received relatively little attention from research workers, they present good opportunities for original work.

Investigation 16: Animals associated with seaweeds

Clumps of seaweed provide cover for a variety of animals, and an extremely rich fauna is usually associated with them. In fact, Colman (1940) has estimated that seaweed faunas may be even richer than those of the soil, which are often cited as examples of exceptional abundance. Comparisons are, of course, difficult, but Colman did it by comparing the numbers of animals per unit area in the two habitats. Whereas the highest figure that he could find for soil samples was 66,747 per square metre, he found an average of 274,320 animals per square metre for rock covered with the lichen, *Lichina pygmaea*.

There are probably many advantages to be gained from living within growths of seaweed. For example, there must be less risk of desiccation among moist fronds than on exposed rocks, and similarly, there are almost certainly smaller fluctuations in temperature. (Actual measurements of the climatic factors within these bunches of seaweed could probably be made; see Cloudsley-Thompson, 1967, for possible ways of doing this.) Seaweeds may also provide some shelter from wave action and may be used as a source of food by some animals.

Most seaweeds are confined to distinct zones on the seashore, and there is also a zonation of the animals associated with them, although just how many of the animal species are plant-specific does not seem to have been studied in any detail. With the notable exception of *Lichina*, which has an unusually large number of animals associated with it (see above), the seaweed fauna high up the shore is normally less abundant than that nearer low water. Periwinkles, *Littorina saxatalis*, and small specimens of the isopod, *Ligia oceanica*, are often common among high-level plants and so is the tiny bivalve, *Lasaea rubra*. Lower down the shore there is usually an increase in the numbers of organisms that are actually attached to the fronds. These include various hydroids, sea-mats and worms, such as *Spirorbis borealis*. But there are also many animals crawling over and among the algae, including the flat periwinkle, *Littorina obtusata*, and the isopod, *Idotea granulosa*.

At low-water mark and below the fauna associated with the oar-weed *Laminaria*, is particularly rich. Dense growths of the sea-mat, *Membranipora membranacea*, and the well-known hydroid, *Obelia geniculata*, are often found on the fronds, and so is the blue-rayed limpet, *Patina pellucida*. The holdfasts of these plants are particularly interesting and by themselves provide a good investigation. The spaces between the rhizoids of the holdfasts are usually filled with detritus, and it is among this that the most extensive fauna is to be found (Plate 7). It includes horse mussels, saddle oysters, hydroids, polyzoans, barnacles, sea squirts, crustaceans and many polychaete worms. There is also an encrusting alga, *Lithopyllum incrustans*, which forms a hard veneer about 1–2 mm thick between the ends of the rhizoids and the rocks to which they are attached.

Seaweeds can be collected from various zones on the seashore. In order to avoid missing many of the organisms that are associated with them, they must be examined carefully, frond by frond. It is also worthwhile to place bunches of algae in dishes of sea-water for a few hours; many animals detach themselves and can be found after the removal of the algae from the dishes. Holdfasts of *Laminaria* can be collected at low tide by pulling the entire plants from the rocks. They should be gripped around the stipe, just above the holdfast, and pulled firmly. Because the animals within them rot rapidly, holdfasts should either be examined immediately or preserved in formalin and examined later. Each one must be broken up piece by piece for examination. Many of the animals in them are small so that a binocular microscope is valuable for their identification, although, even with one, specialist keys may be needed for the identification of some of them to species. Nevertheless, organisms can usually be identified to the correct order or family, which, for many purposes, is adequate. The index in Barrett and Yonge (1958) may also be useful, because under the heading '*Laminaria*' it gives page references to some of the organisms commonly associated with this plant.

References

Barrett, J. and Yonge, C. M. (1958). *Collins Pocket Guide to the Sea Shore*. Collins.

Cloudsley-Thompson, J. L. (1967). *Microecology*. Institute of Biology Series: Studies in Biology. Edward Arnold Ltd.

Colman, J. (1940). On the fauna inhabiting intertidal seaweeds. *J. mar. biol. Ass. U.K.*, 18, 435–476.

Investigation 17: Crevices*

The small crevices and depressions that occur in hard rocks are usually well colonised with animals and, where there is adequate illumination, with plants as well. Such places provide protection from wave action and are usually cool and damp. It is undoubtedly a reflection of the favourable conditions in them that some of their inhabitants (e.g. barnacles) extend further up the shore than elsewhere in the same habitat. They also provide temporary hiding-places for some animals, and are often crammed with periwinkles or dogwhelks when the tide is out.

In soft rocks narrow crevices often occur which are as much as 12 in. deep. These support characteristic populations that are not encountered elsewhere on the shore. So far they have not received much attention, but have been studied in Dartmouth slate at Wembury, Devon, by Morton (1954). Crevices in this rock can be prized open with a crow-bar or some similar tool; motile animals in them can be sucked up in an entomologist's pooter (Figure 13).

The conditions in these crevices vary enormously from one part of the shore to another, depending upon such factors as exposure to the sun, the effect of wave action, the depth of the crevice, the kind of substrate and so on. Nevertheless, there is usually a clear zonation of the crevice faunas from high to low water, as there is on the rest of the shore. In the relatively dry crevices above the high-water mark of spring tides, where there is little organic debris, the tiny periwinkle, *Littorina neritoides*, is often the dominant animal. A little lower down the shore it may be joined by *Littorina saxatalis*, the bivalve, *Lasaea rubra*, and the isopod, *Ligia oceanica*. However, at about high-water mark of neap tides the littorinas tend to drop out, and deposit feeders, such as the worm, *Spirorbis*, and small bivalves (*Lasaea* still occurs), become common. Deposition of silt in crevices is heavier at even lower levels and a strong 'marine' element appears in the fauna in the form of small

* The authors have not had the opportunity to investigate the faunas of crevices in soft rocks, but include the project here because, where suitable rocks exist, it seems to be a particularly interesting one.

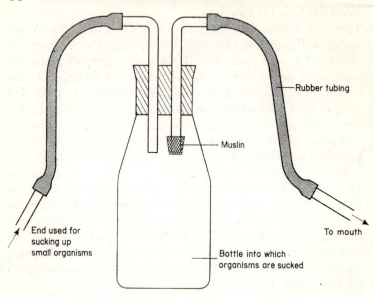

Figure 13. An entomologist's pooter, which can be used for **collecting** small organisms from crevices.

crabs, ophiuroids, saddle oysters and many polychaete worms.

As well as the zonation in crevices from high to low water, individual crevices may also show a zonation which extends from their mouths to the innermost parts. Penetration into the deeper parts of a crevice is associated with several changes, including a decrease in light intensity, less circulation of water when the tide is in and the deposition of more sediment. In response to these changes, the population develops as a series of zones, although all the zones are not represented in every crevice. The full sequence of zones occurs in deep, humid crevices at Wembury (Figure 14; Morton, 1954). The outer zone, around the entrance, is dominated by organisms, such as barnacles and tufts of *Lichina*, which can withstand wave action and are usually common on the shore itself. In the next zone the light intensity is still high enough for the growth of encrusting

Plate 5. The anemone *Adamsia palliata* on the shell of a preserved hermit crab, *Eupagurus pridieauxi*.

Plate 6. Four crabs *Carcinus maenas*. The individuals on the left are infected with the parasite *Sacculina carcini* and those on the right are uninfected. The parasitised male (*upper left*) shows a characteristic broadening of the abdomen compared with the other male (*upper right*). The other two crabs are both females.

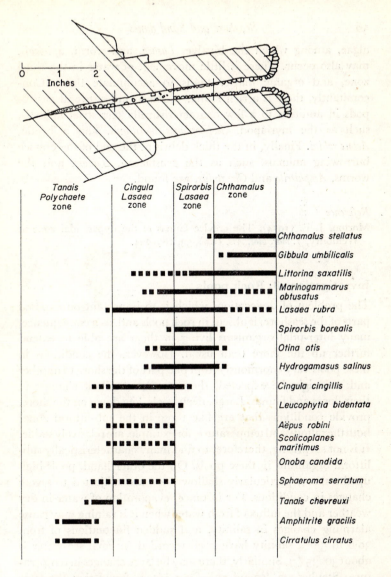

Figure 14. Zonation in a crevice at Wembury Bay, Devon. (Data from Morton; after Lewis.)

E

algae, among which the bivalve, *Lasaea*, and worm, *Spirorbis*, may also occur. There is little movement of water in the third zone, and organic matter tends to accumulate there. Concomitantly, there is a rich fauna, which includes several gastropods in addition to a few characteristically terrestrial animals, such as the myriapod, *Scolicoplanes maritimus*, and a beetle, *Aepus robini*. Finally, in the thick debris at the end of the crevice burrowing animals, such as the crustacean, *Tanais*, and the worms, *Amphitrite* and *Cirratulus*, are found.

Reference

Morton, J. E. (1954). The crevice faunas of the upper tidal zone at Wembury, *J. mar. biol. Ass. U.K.*, 33, 187–224.

Investigation 18: Rock pools

The problem of desiccation, which is so important on exposed parts of the seashore, is absent in rock pools and, as a consequence, many inter-tidal organisms living in them are able to extend further up the shore than usual. However, the conditions in these pools vary enormously from one part of the shore to another and, as might be expected, there are concomitant changes in their floras and faunas. Large, deep pools, low down on the shore, provide conditions that are like those in the sub-littoral zone; both the salinity and temperature, for example, are relatively stable. It is not surprising, therefore, to find many characteristically sub-littoral organisms in these pools. On the other hand, pools high up the shore, particularly shallow ones, are subjected to severe changes in conditions. For instance, evaporation of water in dry weather and the influx of fresh water when it is raining may cause alarming changes in salinity, and sudden fluctuations of from 30·0 to 5·0‰ salinity have been recorded (normal sea-water is about 30–35‰). Similarly, there may be large changes in temperature; Stephenson, Zoond and Eyre (1934) found that the temperature of one high-level pool, which was fully exposed to the sun, rose to a maximum of 13·4° C. above the sea temperature, whereas the temperature of a large low-level pool differed from the sea by only $0.1°$ C. Most marine and inter-tidal animals cannot tolerate such wide fluctuations in conditions, and as one progresses up the shore, they tend to drop out of the rock-pool populations and are replaced by organisms specially adapted to withstand such changes. One of these is the copepod, *Tigriopus fluvus*, which has the remarkable ability to survive in salinities between 4.0 and $90\%_0$. Even salinities above this level do not kill it; it goes into a state of coma, but recovers when the salinity decreases again (Ranade, 1957). There are obvious similarities in the conditions in estuaries and those in high-level rock pools, and it is not surprising, therefore, to find organisms such as the crab, *Carcinus maenas*, the amphipod, *Gammarus deubeni*, and the green alga, *Enteromorpha*, living in both habitats.

(a) *Zonation*

Changes in the inhabitants of rock pools from low to high water can be investigated by making species lists for a series of pools at intervals up the shore, together with estimates of abundancy for each species. The line-transect method is impracticable, because it is unlikely that a series of pools will be found in a straight line or even ones that are similar in depth and extent and, in assessing the results, therefore, it should be remembered that factors other than tidal level may influence the population of each pool. The vertical range of some common organisms, which live both in rock pools and on exposed parts of the shore (e.g. limpets and anemones, *Actinia equina*), can be compared by making abundancy assessments (or actually counting individuals per unit area) in rock pools and areas of the same size on adjacent rocks throughout the tidal range.

(b) *Oxygen content of rock pools*

Changes in the oxygen content in different kinds of rock pools have been investigated by Stephenson, Zoond and Eyre (1934), who found that fluctuations in the content varied both from one pool to another and at different times of day. These researchers recognised three broad categories of rock pool: those with predominantly animal populations (Plate 8), those with predominantly plant populations (Plate 9) and those with mixed populations, and they showed that changes in the oxygen content could be related to the type of pool. For instance, in the 'plant' pools there were large increases in the oxygen concentration during the day; in one case this was from 8·95 mg per litre at 9.45 a.m. to 22·65 mg per litre at 2.00 p.m. There were similar, although less marked increases in the 'mixed' pools, but in the 'animal' pools there were actually decreases in the oxygen content; for example, in one 'animal' pool that was investigated it dropped from 5·1 mg per litre at 8 a.m. to 3·8 mg per litre at 12.30 p.m. These differences can almost certainly be accounted for by the release of oxygen by plants carrying out photosynthesis in the 'plant' and 'mixed' pools during daylight and by the use of oxygen by animals in the 'animal' pools. As would be expected, there is a decrease in the oxygen content in all three types of pool at night. The supply of oxygen is presumably replenished in 'animal' pools when the water is changed by the incoming tide.

Changes in oxygen content can be investigated by collecting samples of water from pools at different times of day. It is usually possible to find typically 'animal' and 'plant' pools on rocky shores. Mixed pools also exist (in reality, all pools are mixed to at least some extent), but unless it is intended to make estimates of the total amount of plant and animal matter in each pool, it may be better to restrict the investigation to the two extremes (i.e. 'plant' and 'animal' pools). In their study, Stephenson, Zoond and Eyre collected and weighed all the plant and animal material. Any gravel in the bottom of a pool was sieved through a $\frac{1}{4}$-inch sieve in order to find the smaller (but not smallest!) animals. They also emptied the pools with a graduated bucket so that the total water content of each could be estimated and the ratio of plant and animal matter to the volume of the pool worked out.

Samples of water for oxygen determination must not be brought into contact with air, and can be collected by submerging bottles of the kind used by Stephenson, Zoond and Eyre in rock pools (Figure 15). These bottles have been specially designed to prevent bubbling as water enters them. The standard method for determining the dissolved oxygen in such samples is that of Winkler. The following reagents are needed:

(*a*) a 40% solution of manganous chloride;

(*b*) a freshly made (not yellow) alkaline iodide solution consisting of 33 g of sodium hydroxide and 10 g of potassium iodide in 100 cm^3 of distilled water;

(*c*) a concentrated non-oxidising acid, such as orthophosphoric acid or sulphuric acid.

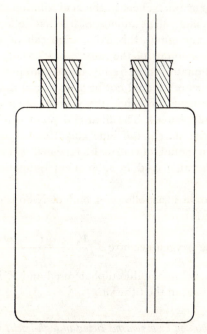

Figure 15. A collecting bottle designed to fill with water without air bubbling through it. The end of the long tube should be submerged so that, as the bottle fills, air is expelled out of the short one.

Samples should be treated with the reagents as soon as possible after drawing (preferably in the field). Add proportionally about 1 cm^3 of the manganous chloride solution to each 150 cm^3 of sample with a pipette whose tip is held well below the surface of the water and then add 0·5 cm^3 of the alkaline iodide solution in the same manner. Stopper the bottle immediately, taking care not to trap any air bubbles, and ignoring any overflow. Shake vigorously, and then leave the precipitate to settle partially before shaking again. This time allow the precipitate to settle almost completely and then add 1 cm^3 of acid; restopper and shake until the precipitate is dissolved.

The oxygen in the original sample is first fixed when it oxidises a proportion of the manganous ions (in the manganous chloride solution) to manganic. Then, in the acid solution, the manganic ions release an equivalent amount of free iodine from the iodide solution. The amount of this free iodine can be estimated by titrating small amounts of the sample, say 50 cm^3, against N/80 sodium thiosulphate, using freshly made starch solution as an indicator. The starch should not be added until the end point of the titration is near (i.e. when the yellow colour of the iodine has almost disappeared). The final end point is reached when, after adding the starch, the blue coloration is just destroyed. This procedure should, of course, be repeated two or three times and the mean value used in subsequent estimation of oxygen content.

The calculation is as follows (1 cm^3 of N/80 is equivalent to 0·1 mg oxygen):

$$\text{mg oxygen per litre} = \frac{V_1 \times 0\cdot1 \times 1000}{V_2}$$

where V_1 is the volume of thiosulphate used and V_2 is the volume of the sample used in the titration.

(c) Osmotic regulation by organisms living in high-level pools

Organisms can be collected from rock pools high up the shore and their ability to osmoregulate successfully tested by keeping them in various dilutions of sea-water (see Investigation 20, page **68**). The salinity of pools can be estimated by titrating samples against standard silver nitrate solution (see page **68**).

References

Ranade, M. R. (1957). Observations of the resistance of *Tigriopus fluvus* (Fischer) to changes in temperature and salinity. *J. mar. biol. Ass. U.K.*, 36, 115–119.

Stephenson, T. A. Zoond, A. and Eyre, J. (1934). The liberation and utilisation of oxygen by the populations of rock pools. *J. exp. Biol.*, 11, 162–172.

5

Life in estuaries

There are marked differences in the floras and faunas of estuaries and those of the seashore, and with the exception of many of the organisms that are characteristic of high-level rock pools, which have much in common with estuaries (see page 56), relatively few species are able to survive in both habitats. The major restricting factor in estuaries is undoubtedly the regular fluctuation in salinity that occurs between one tide and the next. At low water the only inflow into the estuary is from the fresh-water source and, consequently, the salinity is generally low. But as the tide comes in, sea-water mixes with the estuarine water and raises the salinity. These conditions impose severe osmotic problems on organisms that are subjected to them; most of them must suffer an influx of water by osmosis in the dilute sea-water conditions at low tide and, possibly, a loss of water in the more saline conditions at high water. Not many organisms can tolerate such changes, and relatively few species colonise estuaries compared with the numbers that live on the seashore. However, the organisms that have overcome the difficulties of surviving fluctuating salinities often occur in enormous numbers. The burrowing bivalve, *Macoma balthica*, for example, has been recorded at densities of 5900 per square metre in estuarine mud flats.

There appear to be two methods by which animals can withstand fluctuations in salinity. First, the tissues may be adapted to work effectively in a wide range of osmotic pressures so that an influx or loss of water causes no serious harm. The tissues of most estuarine animals are probably capable of tolerating temporary changes of this kind, and there are some, such as the lugworm, *Arenicola*, which survives well in the dilute waters of the Baltic Sea, which can apparently survive permanently reduced salinities. Second, in some animals osmoregulatory mechanisms have been evolved whereby the osmotic flow can actually be regulated. Such a mechanism exists, for example, in the crab, *Carcinus maenas*. If it is placed in dilute sea-water the osmotic pressure of the body fluids falls a little due to the influx of water, but as long as the external medium is not less than 20% sea-water, the concentration is maintained above the lethal limit. It seems that this is done partly by the excretion of water in a dilute urine and partly by the active uptake of salts from the surrounding water against the concentration gradient.

Entry of water into crabs is also limited to those soft parts of the body, such as the gills, which are not covered by the cuticle. However, in the common estuarine worm, *Nereis diversicolor*, there is no such restriction and water enters through any part of the body surface. But this animal can still regulate the osmotic flow, although its ability to do so is less well developed than that of *Carcinus*. Nevertheless, it can survive effectively in dilute sea-water and can actually tolerate lower salinities than *Carcinus*, presumably because its body tissues function better at low osmotic pressures.

Investigation 19: The succession of organisms in estuaries

Mixing of fresh and sea-water in estuaries is a complex process because fresh water is less dense than sea-water and tends to float on top of it, forming a separate layer. In fact, mixing is largely dependent on turbulence, although other factors may be involved, including the state of the tides (e.g. spring or neap) and recent rainfall. Nevertheless, there is invariably a tendency for the water to become less saline as one moves up the estuary, away from the sea, and these changes are usually associated with interesting successions of organisms, particularly in large inlets, where the salinity changes are gradual and may extend for many miles from the sea.

As might be expected, there is a decrease in the numbers of

marine organisms as the water becomes less saline. In an investigation by Bassindale (1941) in the South-West of England, for example, 144 marine animals were collected at Porlock (Somerset), where the water is fully saline, but progressively fewer marine species were found further up the Bristol Channel. At Weston-super-Mare, for instance, there were only about 50, and the numbers diminished still more in the River Severn itself (Figure 16).

This investigation can be repeated by making collections of organisms at intervals along an estuary. Sieves and forks will, of course, be important items of collecting equipment for estuarine mud flats. Good results can be obtained simply by making a species list for each collecting point and then comparing the numbers of marine and fresh-water organisms at each one.

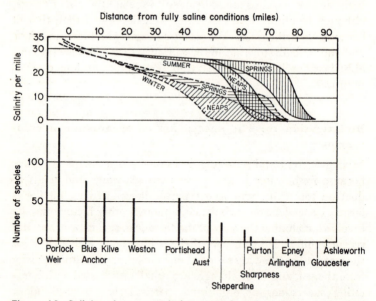

Figure 16. Salinity changes and the approximate numbers of intertidal (marine) animals at a series of stations along the south shore of the Bristol Channel and in the River Severn estuary. (Data from Bassindale; after Yonge.)

Salinities can either be measured with a hydrometer or by titrating samples of water against standard silver nitrate (see page 75). However, the salinity may vary considerably from one part of the estuary to another, even at the same collecting place, so that for good comparative results samples should be taken from the same tidal level and at the same state of the tides (e.g. from low water when the tide is fully out).

There is an interesting succession of species of the amphipod crustacean *Gammarus* in estuaries which is well worth studying. One species replaces another as the water becomes less saline (Figure 17). Near the mouths of estuaries *G. locusta*, a marine form, is usually abundant, but it is replaced in brackish water by *G. zaddachii*. In fact, there are two sub-species of the latter animal: *G. zaddachii zadachii* occurs in the middle reaches of estuaries, but it is replaced in less saline water by *G. zaddachii salinus*. Finally, *G. pulex* becomes the dominant gammarid in fresh water. Other species, *G. chevreuxi* and *G. duebeni*, may also be found, but their distribution is less well known. These species of *Gammarus* provide a particularly good opportunity to study a succession because they are common in most estuaries and can usually be found in large numbers by turning over likely hiding-places, such as stones and pieces of seaweed. Furthermore,

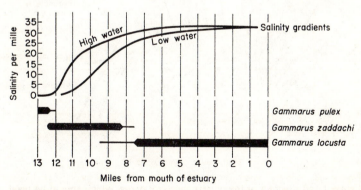

Figure 17. Zonation of the crustacean *Gammarus* in an estuary. (Data from Serventy; after Yonge.)

gammarids seem to be largely indifferent to changes in the nature of the substratum, so that there is usually a continuous population of them along the estuary, even though the species change from one end to the other. They do not normally show the patchiness in distribution that is a feature of some organisms.

Collecting *Gammarus* is made easier if a box sieve is used. This should be about a foot square with a bottom of perforated zinc. Gravel, stones, bits of seaweed and other objects that are likely to harbour *Gammarus* should be placed in the sieve, which is then shaken in water. At one time identification of species was difficult, but thanks to work by Sexton (1942) and Spooner (1947), it is now an easier task. However, details of species are not given in the commonly used books on the seashore, and a key for their identification is given below. It should be used in conjunction with Figure 18. Two common amphipods, *Melita* and *Marinogammarus*, which might otherwise be confused with *Gammarus* itself are also included. *The key includes only estuarine species.*

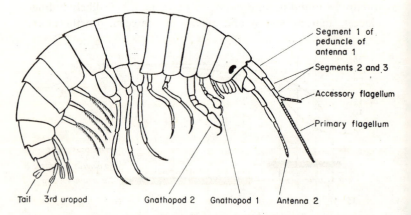

Figure 18. A diagram of the crustacean *Gammarus*, which should be used in conjunction with the key in the text for the identification of different species.

Key for the identification of estuarine species of *Gammarus*

1. Spines on the last three segments of abdomen and tail . 2
   No spines on the last three segments of abdomen and tail
   Melita spp.

2. Inner branch of the last pair of appendages (third uropod) less than half the length of the outer branch *Marinogammarus* spp.
   Inner branch of the last pair of appendages more than half the length of the outer branch 3

3. The hand of gnathopod 1 is smaller than the hand of gnathopod 2; the primary flagellum has 34 or more joints (unless damaged), and the accessory flagellum is more than 7 jointed. 4
   The hand of gnathopod 1 is about the same size as the hand of gnathopod 2; the primary flagellum is less than 34 jointed and the accessory flagellum less than 7 jointed . . . 6

4. Segment 1 of the peduncle of antenna 1 is about as long as segments 2 and 3 of antenna 1 combined; the accessory flagellum has 8–14 joints. Antenna 2 has 24 joints . *G. locusta*
   Segment 1 of the peduncle of antenna 1 is only slightly more than half as long as segments 2 and 3 combined; the accessory flagellum has 7–9 joints. Antenna 2 has 15 joints . . 5

5. Appendages, tail and antennae with dense, long hairs and setae . . . . . . *G. zaddachii zaddachii*
   Appendages, tail and antennae with fewer, shorter hairs and setae . . . . . . *G. zaddachii salinus*

6. Small eyes; the flagellum of antenna 2 has 13 segments
   G. pulex
   Large eyes; the flagellum of antenna 2 has less than 10 segments 7

7. Antenna 2 is almost as long as antenna 1; the 4th and 5th segments of antenna 2 are about the same length as one another . . . . *G. duebeni*
   Antenna 2 is much shorter than antenna 1 (ratio about 2:3); the 5th segment of antenna 2 is shorter than the 4th segment
   G. chevreuxi

References

Bassindale, R. (1941). Studies on the biology of the Bristol Channel. IV. The invertebrate fauna of the southern shores of the Bristol Channel and Severn Estuary. *Proc. Bristol Nat. Soc.*, 9, 143–201.

Bassindale, R. (1942). Studies on the biology of the Bristol Channel. VII. The distribution of amphipods in the Severn Estuary and Bristol Channel. *J. Anim. Ecol.*, 11, 131–144.

Sexton, E. W. (1942). The relation of *Gammarus zaddachii* Sexton to some other species of *Gammarus* occurring in fresh, estuarine and marine waters. *J. mar. biol. Ass. U.K.*, 25, 575–606.

Spooner, G. M. (1947). The distribution of *Gammarus* species in estuaries. *J. mar. biol. Ass. U.K.*, 27, 1–52.

Investigation 20: Osmotic regulation in *Nereis diversicolor*, *Carcinus maenas*, *Gammarus* spp. and other animals

The ability of the ragworm, *Nereis diversicolor*, to osmoregulate in dilute sea-water can be investigated by using changes in weight of the animal as an index of the amount of water loss or uptake. Individuals, adapted to living in 100% sea-water by keeping them in it for about a day, should be weighed, after dabbing them with absorbent paper to remove excess moisture, and then transferred to water of lower salinity (e.g. 50%). Subsequent weighings at fifteen-minute intervals should reveal an increase of weight due to the influx of water. However, the gain in weight becomes less marked, as the influx is controlled, and eventually there may actually be a decrease in weight, presumably because the worm excretes excess water. In this experiment it is interesting to compare the ability to osmoregulate in the estuarine species, *N. diversicolor* (*N. virens* also occurs in estuaries), and one of the characteristic species of the seashore, such as *N. pelagica* or *Perinereis cultrifera*.

Unfortunately, changes in weight in crabs, *Carcinus maenas*, are unreliable guides to osmotic uptake or loss, because water is held in variable amounts in the gill chamber, which itself leads to considerable fluctuations in weight. However, the ability of crabs to survive in different dilutions of sea-water can be used as an indication of their ability to osmoregulate successfully. Crabs should be left in various dilutions (e.g. 0%; 20%; 40%; 60%; 80%; and 100% sea-water in distilled water) for at least a day and then examined so that the mortality rate can be assessed for each dilution. Further experiments, over a narrower range of salinities, will enable more exact estimates of the salinity tolerance to be made. Other animals can be used as subjects in this experiment. It is, for example, interesting to work out the salinity tolerances of different species of *Gammarus*, particularly if their distribution in an estuary has been worked out beforehand.

6

Sand dunes

Strictly speaking, sand dunes are not part of the littoral zone, but they provide an excellent opportunity for investigations of the changes associated with plant successions and are well worthy of attention. Their development from hillocks of shifting sand near the drift line, which are colonised by only a few specialised plants, to a habitat supporting a mature terrestrial flora is a slow process taking many hundreds of years. However, because they are formed as a series of ridges parallel to the sea, with the youngest ones nearest the shore and progressively older ones replacing them inland, several stages in the succession exist side by side at the same time.

Sand dunes first develop when sand, which is blown from the seashore, meets obstacles in the drift line, such as dead seaweed and drift wood, and accumulates around them, forming slightly raised hillocks. Plants, such as the sand couch grass, *Agropyron junceiforme*, and the sea rocket, *Cakile maritima*, can grow in these embryo dunes, and more sand accumulates around them, further increasing the sizes of the hillocks. The next stage in dune development, the so-called yellow dune stage, is characterised by the growth of marram grass, *Ammophila arenaria*, which produces numerous underground rhizomes. The accumulation of blown sand continues, and in many places the yellow dunes reach enormous heights of 50 ft. or more. In the grey dune stage, which succeeds the yellow dune, the stabilisation of sand allows the invasion of a variety of new plants, and complete vegetation cover occurs for the first time. Marram grass does not survive well in them (it appears that its growth is actually stimulated by moving sand), and it is replaced by other grasses, usually by creeping fescue, *Festuca rubra* var. *arenaria*, and the sand sedge,

Plate 7. A holdfast of the oar-weed *Laminaria*.

Plate 8 (*over right*). A rock pool dominated by animals, mainly periwinkles and limpets.
Plate 9 (*over left*). A rock pool dominated by red and brown algae.

Carex arenaria. Many non-maritime species colonise this stage, including angiosperms, such as bird's-foot trefoil, *Lotus corniculatus*, and ragwort, *Senecio jacobaea*, as well as several mosses and lichens.

Subsequent stages are influenced by soil characteristics. Usually dune soil tends to be acidic, and an acid heath vegetation develops on it. But in places where the calcium content of the soil is high a calcareous heathland may develop instead. Both of these may eventually be replaced by scrub- and then wood-land.

The successive (seral) changes in the development of sand dunes can be investigated by making transects from the drift line on the seashore to the grey dunes and beyond. Basically, the method suggested for transects on the seashore (see page 3) can be used. Semi-permanent transects can be made with pegs and string, which has been marked with indian ink at intervals of 5, 10 or more metres, depending on the extent of the dunes being studied and the time available. These marks can be used to indicate successive stations along the transect. At each one the quadrat method can be used to estimate the plants present. A wire (quadrat) frame (1 m^2 is a suitable size) should be thrown down and the plants within it identified. The abundancy of each species can be assessed by counting the numbers of specimens within the frame.

Some changes in the dunes, which can be determined by simple measurements, are characteristic of successions in general. For example, there is usually an increase in the numbers of species present as the habitat ages and, similarly, an increase in the density (numbers per unit area) of the population. The total mass of dead and living organic matter also increases with age. Rough estimates of the total organic matter at different stages of dune formation can be made by determining the humus content (= dead organic matter) for a given area (say 1 m^2) and then the total dry weight of the living matter in the same area. The method for humus determination is given below (page 75); living matter can be estimated after denuding the area of all

F

animals and plants (complete with roots), which should be oven dried and weighed.

Investigation 21: Changes in the soil during dune development

The soil changes considerably during the development of dunes, and there is little doubt that this is one of the most important factors affecting the growth of plants in them. A variety of different processes are involved in its formation. One of the most important of these is the production of humus, which provides a source of nutrients for plant growth as well as binding sand grains together and thereby helping to stabilise the soil. As might be expected, the early dune stages are very poor in humus, although it increases in the later stages (Figure 19).

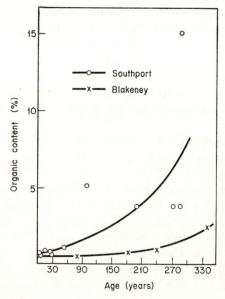

Figure 19. Changes in the organic (humus) content of sand dunes at Southport and Blakeney Point, Norfolk, with ageing. (After Salisbury.)

Humus content also influences the soil pH, and with age there is a tendency for dunes to become more acidic (Table 4). However, the presence of carbonates, resulting from the weathering of the shells of molluscs, sometimes counteracts the influence of humus, and some dune soils give an alkaline reaction. But this is unusual. Even where carbonates are present, they are normally leached down to the lower layers, so that the surface soil may be almost totally lacking in them. Leaching also leads to a progressive decrease in the carbonate content as dunes age; in Salisbury's study, for instance, there was a fall from 4% carbonate in the drift line to 0·42% in the embryo dunes and 0·34% in the yellow dunes.

The water-holding capacity of the soil also increases as more humus is formed, and this, too, is reflected in an increase in water content as dunes age. But the water content of dune soil is never very large, particularly in the early stages, where there is little humus to improve the water-retaining properties. However, experiments by Salisbury (1922) suggest that it is more readily available than is sometimes supposed. He found that, although typical dune plants, such as Hare's foot, *Trifolium arvense*, could exhaust the entire water content of the sand in which their roots were distributed in a few days, healthy plants could still be found in the dunes after as much as six weeks without

TABLE 4

The variation of pH with age in sand dunes at South Haven (Dorset). (After Chapman.)

pH	Dune type	Dune age (in years)
6·6–7·0	Early dune	0– 20
5·0–5·5	Dune grassland	0– 50
4·8–5·5	Late dune grassland	50– 80
3·9–4·6	Dune heath	80–110
3·9–4·5	Dune heath	110–230
3·6–4·5	Dry heath	240–350

rain. Clearly these plants must use some source of water other than that from rainfall. The explanation appears to be that dew is regularly deposited in the dune soil, and this replaces the water lost by plants. Heat loss from sand is extremely rapid, and even in mid-summer, the internal temperature of the dunes falls below the dew point at night, and dew is steadily deposited in them.

There is less salt than might be expected in all but the embryo dunes. In these salt content is a critical factor for plant growth, and only plants, such as *Agropyron junceiforme*, which can tolerate high salinities can survive in them (Salisbury, 1952). Even marram grass, which can live well in shifting sand, cannot colonise embryo dunes because of their high salt content. There is less salt in older dunes, probably because it leaches out of them, and it no longer seems to be a limiting factor.

Soil characteristics can be investigated by taking samples of soil from each of the major stages and analysing them in the laboratory. Embryo dunes are not always present, even where older stages exist, but if they are not, samples can be taken from the drift line. The soil can be taken from root level at each stage; if samples are kept in closed polythene bags there is no water loss from them during transportation.

(a) Water content

This can be estimated from the loss of weight from samples when they are dried. Samples (say about 20 g) should be weighed and then oven dried at about 40° C. High temperatures (e.g. from heating with a bunsen burner) must be avoided, because there is a danger of burning the humus present. Each sample should be reweighed at intervals until there is no further loss in weight. The percentage loss in fresh weight, which is equal to the amount of water present, can then be calculated. The effect of internal dew formation on the water content can be shown by analysing samples of soil collected in the evening, before dew deposition, and in the early morning after it.

(*b*) *Humus content*

The humus in a sample of soil can be estimated by oxidising that present with hydrogen peroxide (20 volume) and then measuring the loss in weight that occurs. Burning (calcining) the humus is inaccurate if carbonates are present in the soil.

The recommended method is to oven-dry (at about 40° C.) samples of soil (about 3 g), which are then weighed accurately and placed in beakers. 20 cm³ of hydrogen peroxide is added to each of them and the resulting mixture heated gently until effervescence has ceased. Then the samples are heated again until they become light coloured and there is no further colour change. After cooling they are filtered and washed with warm distilled water. Each sample must then be oven dried and weighed. The loss in weight will represent the humus originally present in the sample.

(*c*) *Salt content*

The method used by Chapman (1939) depends upon titrating the salt (sodium chloride) against standard silver nitrate.

Using this method 10-g samples of air-dried soil are placed in flasks containing 100 cm³ of 1% sodium sulphate solution (this facilitates the subsequent filtration). This is left for about 24 hours and then the solution is filtered and the sand in the filter washed with 50 cm³ of distilled water to dissolve any chloride that remains in it. This results in a filtered solution of 150 cm³ containing the sodium chloride present in the original sample; if the amounts suggested above are used, 15 cm³ of solution will contain the amount of chloride in 1 g of the sample. This is titrated against standard silver nitrate solution using potassium chromate as an indicator. The percentage of sodium chloride present is then a straightforward calculation; 1 cm³ standard silver nitrate is equivalent to 0·001 g chloride. (Estimations of the chloride content of sea-water can be made by titrating samples against standard silver nitrate solution.)

(d) *pH*

This can be estimated by using universal indicator. Small samples of soil should be placed in test tubes and a few cm³ of distilled water added. The supernatant liquid should then be tested with universal indicator.

(e) *Carbonate content*

Carbonate content can be estimated indirectly by an acid–alkali back titration. It is said to be accurate to the nearest 1%.

The method recommended by Barnes (1959) is as follows: 5 g of the soil sample is transferred to a tall 150-cm³ beaker and 100 cm³ of N hydrochloric acid added to it. The contents are stirred vigorously during the period of an hour; between stirrings, the beaker is covered with a watch glass. Then, after the mixture has settled, 20 cm³ of the supernatant liquid is transferred to a titration flask. About 6–8 drops of bromothymol blue are added as an indicator before it is titrated against N-sodium hydroxide. (It may be necessary to add more indicator if the colour tends to fade near the end point of the titration.) Finally, a blank titration (N-HCl against N-NaOH) is carried out. The calculation is then as follows:

$$\text{Percentage of carbonate} = (V_1 - V_2) \times 5$$

where V_1 is the volume of hydrochloric acid needed to neutralise the sodium hydroxide in the blank titration and V_2 is the volume needed to neutralise the sodium hydroxide in the first titration.

References

Barnes, H. (1959). *Apparatus and Methods of Oceanography*. Allen and Unwin.

Chapman, V. J. (1939). Studies in Salt Marsh Ecology. Sections IV and V. *J. Ecol.*, 27, 160–201.

Salisbury, E. J. (1922). The soils of Blakeney Point: a study of soil reaction and succession in relation to plant covering. *Ann. Bot.*, 36, 391–431.

Salisbury, E. J. (1925). Note on the edaphic succession in some dune soils with special reference to the time factor. *J. Ecol.*, 13, 322–328.

Salisbury, E. J. (1952). *Downs and Dunes*. Bell.

Investigation 22: Special adaptations of some dune plants

The peculiarly demanding nature of sand dunes as a habitat suitable for plant growth has led to several special adaptations. On the grey dunes, for example, there are many annuals, such as the Mouse-ear chickweeds, *Cerastium tetrandrum* and *C. semi-decandrum*, which pass through their life cycles in winter when moisture is more plentiful.

Plants inhabiting dunes are nearly all xeromorphic, and many have special adaptations of the leaves which reduce water loss by transpiration. These include hairy leaves, spined leaves and thick cuticles, which probably also serve to prevent injury from the impact of sand grains blown by the wind. These adaptations can be studied by cutting sections of the leaves and examining them microscopically.

The curled leaves of marram grass are particularly interesting (Figure 20). The upper (outer) epidermis has a thick cuticle, while the lower epidermis, on which the stomatal pores are

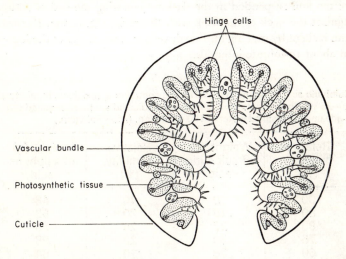

Figure 20. A diagrammatic section of a leaf of marram grass, *Ammophila arenaria*. (After Fritsch and Salisbury.)

located, has a thin cuticle and is in a protected position on the inside. In fact, the inner surface of the leaf has a corrugated appearance, with the stomata confined to the furrows, which are lined with interlocking hairs. Special, large hinge cells at the base of these furrows are the first to lose water when the supply to the leaf becomes short. When they are turgid, pressure from them on the surrounding tissues tends to hold the furrows open so that the leaf is relatively flat and the inner epidermis is exposed to the air. However, when the hinge cells lose water and become flaccid the furrows close so that the leaf curls and becomes roughly circular in section. This habit almost certainly reduces the rate of water loss from the inner epidermis in conditions of drought.

The curling habit of *Ammophila* leaves can be investigated by suspending cut pieces of leaf in a series of conical flasks or similar containers in which the relative humidity is fixed at different values (at R.H. values of, say, 100, 75, 50, 25, 5) by solutions of sulphuric acid of various concentrations (Table 5). Lengths of leaf, about 2 cm long, should be threaded with needle and cotton and suspended in the flask by jamming the end of cotton against the neck of the flask with the cork (Figure 21). Records can be kept by making outline drawings of the stages of the leaves at about 10-minute intervals.

TABLE 5

Sulphuric acid solutions in water for the control of atmospheric humidity at 20–25° C. The weights of sulphuric acid indicated are those required to be made up with distilled water to 100 g solution. (After Solomon.)

Relative humidity (%)	Weight (g) of sulphuric acid	Relative humidity (%)	Weight (g) of sulphuric acid
100	0	40	48
90	18	30	52
80	27	25	55
75	30	20	58
60	38	10	64
50	43	5	70

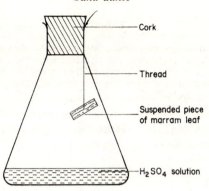

Figure 21. The suggested arrangement for suspending pieces of marram-grass leaf in atmospheres of fixed humidity.

Investigation 23: the 'tussock' growth habbit in marram grass

Aerial shoots of marram grass usually aggregate in groups to form the 'tussocks', which are a characteristic feature of yellow dunes. This growth habit may be advantageous to *Ammophila*, because the dense tussocks form obstacles to the wind and, as a result, help to stabilise the sand. Blown sand accumulates around them, forming raised hillocks.

The aerial shoots develop from the axillary buds of underground rhizomes which, in this grass, may be several feet in length. Gimmingham, Gemmell and Greig-Smith (1948), who investigated the growth habit of marram grass in the Outer Hebrides, suggested that a tussock forms because, once one aerial shoot develops above the surface of the ground and grows successfully, it enhances the food reserves of the rhizome below it, and thereby stimulates the growth of nearby buds. Consequently, several shoots appear within a limited area. This hypothesis was supported by the examination of rhizomes because it was found that aerial shoots tended to occur in groups along them; axillary buds of several successive nodes gave rise to shoots, while series of buds on either side remained undeveloped (Tables 6 and 7).

TABLE 6

The observed numbers of consecutive nodes of *Ammophila* rhizomes bearing leafy shoots compared with numbers expected by chance. There are fewer singly occurring leafy shoots than expected but more large groups of shoots (more than four) occurring consecutively. (Data from Gimmingham, Gemell and Greig-Smith.)

Number of consecutive nodes bearing leafy shoots	Observed numbers	Expected numbers (to nearest 0·1)
1	15	48·9
2	6	12·9
3	3	3·4
4	2	0·9
6	3	0·1
8	2	0
12	1	0

Furthermore, shoots at the centre of each group tended to be longer than those at the ends, suggesting that they had started growth at an earlier date.

Gimmingham *et al.* also found that there tended to be a cyclic variation in the lengths of internodes along the rhizomes examined by them. That is to say, there were alternations of progressively longer internodes with progressively shorter ones (Table 7). There was also a tendency for aerial shoots to emerge from the buds at the ends of relatively long internodes. These workers suggest that long internodes will have greater food reserves, and will therefore be better able to support the growth of an aerial shoot.

Rhizomes for examination should be carefully excavated with trowels and forks. Each should be kept as long as possible; short, broken pieces of rhizome are, of course, unsuitable. The distribution of leafy shoots along each rhizome should be recorded in the manner shown in Table 7 and the lengths of shoots measured so that those at the ends of groups can be compared

TABLE 7

The occurrence of leafy shoots (LS) and the inter-node lengths along three of the rhizomes examined by Gimmingham, Gemmel and Grieg-Smith.

Plant A		Plant B		Plant C	
2·7	2·1	3·7	4·5 LS	0·7	4·9 LS
4·0	3·0 LS	5·0	4·7	1·0	4·5
3·7	3·9 LS	5·0	4·1 LS	6·2	4·9
2·9	6·0 LS	4·3	2·6 LS	8·0	4·8
1·8	6·5 LS	4·9	2·9	7·2	3·3
1·9	2·9 LS	5·1	4·0	6·8	1·6
2·4	1·7 LS	6·0	3·7	5·9 LS	3·5
2·6	2·9	4·8	3·2	5·9 LS	4·5
2·2	3·4 LS	5·0	3·4	5·2 LS	4·7
2·1	2·5 LS	5·7	2·2	6·3 LS	3·2
2·1	Apex	6·3	1·9	5·4 LS	
2·0		6·2	3·2		
		5·5 LS	2·1		
		6·5 LS	2·6		
		7·0 LS	3·7		
		5·7 LS	3·0		
		5·6 LS	2·8		
		6·2 LS	Apex		
		5·0 LS			

with those in the middle. Internode lengths should be measured and the data examined for evidence of cyclic variation. The mean length of internodes bearing shoots can be compared with that of internodes not bearing them.

Reference
Gimmingham, C. H., Gemmell, A. R. and Greig-Smith, P. (1947). Tussock formation in *Ammophila arenaria* (L.) Link. *New Phyt.*, 46 262–268.

Index